D0385185

THE **TINKERERS**

THE **TINKERERS**

The Amateurs, DIYers, *and* Inventors
WHO MAKE AMERICA GREAT

Alec Foege

BASIC
BOOKS
A Member of the Perseus Books Group
New York

Books published by Basic Books are available at special discounts
for bulk purchases in the United States by corporations, institutions,
and other organizations. For more information, please contact the Special
Markets Department at the Perseus Books Group, 2300 Chestnut
Street, Suite 200, Philadelphia, PA 19103, or call (800) 810-4145,
ext. 5000, or e-mail special.markets@perseusbooks.com.

Book designed by Linda Mark
Set in 10 pt Berkeley Oldstyle

Library of Congress Cataloging-in-Publication Data

Foege, Alec.
  The tinkerers : the amateurs, DIYers, and inventors
who make America great / Alec Foege.
    p. cm.
  Includes bibliographical references and index.
  ISBN 978-0-465-00923-7 (hardcover : alk. paper)—
  ISBN 978-0-465-03345-4 (e-book)
  1. Tinkers—United States. 2. Inventors—United
States. I. Title.

  HD8039.T572U64 2013
   609.2'273—dc23
                                    2012028740

10 9 8 7 6 5 4 3 2 1

For my wife, Erica, who knows a thing or two about ingenuity

CONTENTS

CONTENTS

CHAPTER 1

WISING UP ABOUT A SMARTPHONE

A FEW YEARS AGO I ENGAGED my then two-month-old smartphone, a BlackBerry of some sort or another, in a very nontechnical road test: I sat on it. I only noticed the damage when one afternoon I reached to check my email. The small screen, usually jittering and scrolling with plenty of new messages, was suddenly a disconcerting Technicolor swirl with a huge black spot in the middle. A Rorschach test for the addled info junkie.

Suffering from the withdrawal, symptoms familiar to anyone addicted to their phone, I drove in a mild panic to the nearest Verizon Wireless store, located in a small strip mall in a neighboring town.

After a short wait, I met with a sales representative seated in front of a computer screen. After asking for my vitals, he typed for a few seconds and waited. Then he typed, then he waited. Then he sighed.

"You can get a new phone," he said.

"Free of charge?" I said, already knowing the answer.

"No," he responded. "At retail price."

"How much is that?" I asked.

"Four hundred fifty dollars."

Could I get my current BlackBerry fixed? The rep shook his head sadly. "They don't let us repair the phones in the store anymore," he said. "That was my favorite part of the job. Now all I get to do is sell phones."

I felt his pain. Having grown up tinkering with Radio Shack electronic kits, I used to love taking things apart—radios, tape players, anything I could get my hands on.

But in the last twenty-five years or so, the number of household devices we can easily tinker with has dwindled.

When I arrived home, I dug out my old BlackBerry. Two and a half years earlier, I had marveled at its slick design and state-of-the-art "world phone" capability. Now it just looked thick and clunky. And what would I do without its previously special ability to make calls from other countries without swapping out a computer chip? It didn't matter since virtually every phone can do that now.

I googled my model number to see if I could find a more affordable replacement. What I stumbled onto instead was a short video on YouTube. The video showed a pair of hands disassembling a BlackBerry and replacing the screen in a matter of minutes. A male voice, with an appealingly clipped English accent, guided me through each step.

I was hooked.

Through another Google search, I found an online retailer selling replacement screens for around $45, as well as a small smartphone-specific toolkit, including a tiny torque screwdriver and a little plastic tool for prying apart the BlackBerry's flimsy case. One FedEx delivery later, I had my phone disassembled and its parts neatly laid out on my desk. The screws came out easily; the case popped right off. Inside the phone, there were remarkably few parts. Following the YouTube video instructions carefully, I was able to unplug the broken screen, which was attached to the circuit board by a tapelike digital connector leading to a six-pin plug. I snapped in the new screen with little trouble, clicked the case back together, and tightened up the tiny screws with my tiny torque screwdriver.

Just ten minutes after starting the process, I powered it up. Good as new.

My tinkering journey ended at the point when I had a working phone again. But it certainly didn't have to. Having discovered through my own persistence that this modern-age bit of machinery wasn't quite as complicated as I had first thought, I might have been emboldened to make my own alterations to it.

Indeed, a quick online search revealed the fruits of a few intrepid BlackBerry tinkerers. One was titled "How to Convert a BlackBerry Camera into a Webcam." Another demonstrated how to reverse-engineer a BlackBerry into a complete home automation control system.

Perhaps the best example of the smartphone-tinkering phenomenon is the remarkable case of George Hotz. Hotz came to fame in 2007 as a seventeen-year-old hacker of Apple's iPhone.

Hotz, a T-Mobile subscriber, wanted to use the iPhone with his existing phone plan. But at the time, Apple had an exclusive deal with AT&T. Armed with nothing more than an eyeglass screwdriver, a guitar pick, and a soldering gun, he was able to erase his iPhone's baseband processor, the computer chip that determined which phone carriers the device would operate with. On his own PC, he wrote a new string of code for his iPhone, allowing it to operate with any wireless network. Hotz staked his claim as the first person to unlock an iPhone. This accomplishment quickly brought him both fame and notoriety.

A few years later, in January 2010, Hotz succeeded in unlocking a Sony PlayStation 3 video-game console, which ignited a torrent of malfeasant hacking, culminating in a grand attack by a hacking group known as Anonymous that temporarily forced Sony to shut down its PS3 online gaming network.

I don't mention Hotz's story as evidence of hackers wreaking havoc, but rather to show the immediate power seemingly innocuous tinkering can have in contemporary society. It's important to note here that Hotz viewed himself as performing a valuable service to society in both of these cases.

And Hotz's impressive resume as a tinkerer backed up his claim. While still in high school, he invented a personal transportation device called the Neuropilot that users could drive around just by thinking about it. His senior year, he won a $15,000 science-fair prize for building a 3-D display. In May 2011, Sony extended an invitation to Hotz to visit its American headquarters, where he met with engineers working on the PS3 and explained how he broke into their system.

Where do we draw the line between tinkerers and hackers? What role does tinkering play in contemporary society? How did tinkerers traditionally influence American industry and society? Do we still have what it takes as a nation of tinkerers to excel in the global economy? This book explores the impact American tinkerers have had on the growth of the nation, and what role they may play in our future. It also explores some of the cutting-edge approaches being taken to address what some fear is the waning American tinkering spirit.

I believe the answers to these questions lie somewhere in the tension between corporate discipline and individual ingenuity. My experience with my BlackBerry is a perfect example. With the rapid decline of Research In Motion—the company that manufactured it—since I first purchased it, no doubt new ideas for repurposing these smartphones are cropping up every day.

But there is no guarantee that the best ideas will ever be realized, much less filter into the marketplace. That's because too many average people are discouraged from ever opening these gadgets and examining how they work. Of course, large manufacturers would prefer that we simply toss them out and replace them with something shiny and new. That's just the nature of capitalism.

But there are some fresh avenues emerging through which the United States as a leading economy and culture can improve the odds that the finest work of its most talented tinkerers finds its way into the commercial mainstream. This book explores some of those avenues.

The clerk's sadness at not being able to fix things anymore, and my own sense of accomplishment at having avoided the "throw out and replace" syndrome, I think, were both symptomatic of something afoot in our culture: a return to an important tradition that has been to some extent a casualty of the remarkable efficiency with which we produce all manner of stuff. For many generations in the postindustrial age, puttering around with the mechanical devices that surrounded us was practically a rite of passage, and for many, a way of life. It tethered us to our machines and reaffirmed our notions of modern civilization. Deeply probing how things worked also provided children and adults alike with endless hours of enjoyment. It saved enterprising souls hundreds if not thousands of dollars on repair bills. It also often resulted in new and startling discoveries that sometimes led to fresh innovations.

The first gadget Steve Jobs cobbled together while still in high school with his geeky older college buddy Steve Wozniak was a "blue box" that enabled free long-distance phone calls by duplicating the appropriate digital tones. Sure, the blue box was illegal, but that didn't stop the mismatched pair of "phone phreaks" from selling a bunch of the units to college students and other intrepid pranksters.

The blue box grew out of a simple love for playing around with gadgets and making them bend to the will of a few individuals.

Jobs and Woz, who later cofounded Apple Computers, weren't preparing for a career in hacking phone service. Rather they were engaging in the time-tested American tradition of tinkering with what was around them, and through doing so, exploring their potential for future innovation. It didn't hurt that Jobs had grown up next door to a Hewlett-Packard engineer who liked to tinker with electronics in his garage and who let him watch. Or that he became a member of the Hewlett-Packard Explorer Club as a teenager, where he was exposed to the company's new inventions in an up-close fashion.

The word "tinkerer" had, until recently, a slightly negative connotation, suggesting individuals who are somehow aimless, lacking focus, or not sufficiently motivated to create something genuinely new. To many, "tinkering" sounds like a quaint pastime reserved for those who are retired or otherwise disengaged from the everyday process of mainstream productivity. That's if they think of tinkerers at all. The term itself has fallen out of use, at least in the traditional sense. But historically, American tinkerers were a relatively eclectic bunch who hatched extraordinary, life-changing innovations by sheer will and forward momentum. Benjamin Franklin, Eli Whitney, Cyrus Mc-Cormick, Samuel Morse, Charles Goodyear, Thomas Edison, the Wright Brothers.

Then life got more complicated. It's often assumed that somewhere in the late 1800s, at the turn of the century, tinkerers went the way of the horse and buggy. But I would argue to the contrary: America's tinkering tradition has always been a key part of its ongoing greatness.

So what do I mean by tinkering, in contemporary terms? At its most basic level, tinkering is making something genuinely new out of the things that already surround us. Secondly, tinkering is something that happens without an initial sense of purpose, or at least with a purpose quite different from the one originally identified. Tinkering also emanates from a place of passion or obsession. Lastly, tinkering is a disruptive act in which the tinkerer pivots from history and begins a new journey that results in innovation, invention, and illumination.

Increasingly, however, American tinkering is the unlikely by-product of a country driven by greed and conformity. Within our success as a nation and a global economy lies a paradox. The United States, with its highly disciplined approach to capitalism, invented the modern corporation and the marvelous, sleek objects it produces. Indeed, our processes have become so rationalized and efficient that we can produce new things that are cheaper than the old ones they replace. But as those wondrous corporations become bigger and more efficient, they conspire to take control of many of the outlets of our tinkering, threatening to snuff out the very creativity and brilliance that fueled the growth of those corporations. Still, American tinkering nonetheless prevails: prosperity may have made many Americans fat and happy, but it also gave other Americans just enough leisure time to pursue that which seems almost futile.

What are the characteristics of tinkerers? They are smart and immersed in the world but not necessarily trained in a specific field. They may be affiliated with large corporations or institutions, but they rarely fit in due to their desire to pursue their own interests. They are generalists in a world of specialists. They might be inventors, but they don't necessarily have to be. Mere inventors set out with an assigned goal, such as devising an electric car with enough power and range to supplant gas-fueled ones. They even can be trained engineers with a penchant for unstructured exploration. Tinkerers are focused on fiddling around with what they find around them, and in the process taking existing inventions and repurposing them, as in the case of George Hotz, and in some instances solving problems that the culture doesn't even know need solving.

In other words, tinkerers can be anyone with big ideas and the time to pursue them.

Tinkerers may not have a clear goal, but that's what makes them so exciting and dynamic to the culture around them. Tinkerers are dilettantes, but in a good way. They are optimists with the mental fortitude to shape their optimism into something concrete.

Messing with the innards of a BlackBerry isn't, as it turns out, as fraught with difficulty as you might have imagined. And using a video with step-by-step instructions to fix something isn't quite the same thing as disassembling it and figuring it out yourself. But the willingness to try and the refusal to be cowed by the powers that be (in this case the manufacturer that implores users not to break into their products and—the horror—void the warranty) is something intrinsically American. This is not to say that it doesn't exist in other cultures, but rather that Americans imbue it with a unique mix of cockeyed optimism and brash madness. In this context, middling acts and muddling and puttering and tinkering can become something noble, even transforming, in the right hands at the right moment and with the right problems to solve.

Once upon a time, the United States was a nation of tinkerers, both formally trained and homespun innovators who solved the nation's biggest problems, mostly from behind the scenes. Now, after an era of economic excess that transformed our nation from one of doers to consumers, the United States risks losing its hallowed tinkerer tradition—as well as the engine of innovation that fueled an unprecedented era of growth. Economic success has given us the time and resources to tinker, but it has also blunted our impetus to do so.

A National Science Board report released in May 2010 noted that US investment in research and development has remained essentially flat since the 1980s, at 2.7 percent, though the federal government's contribution has steadily declined while governments of countries such as Japan and South Korea have increased their scientific funding.

More astonishingly, in our technological age, only one-third of American college students earn degrees in science or engineering. While trained engineers have no monopoly on professional tinkering, they tend to be less intimidated by our modern-day gadgetry. The comparative figures are 63 percent in Japan and 53 percent in China. For a long time, the United States ranked among the countries with the highest ratio of engineering and science degrees; now we're near the end of the list of twenty-four countries that track such data. And the economic growth that once went hand in hand with that big-sky imagining and doing is in jeopardy of moving elsewhere.

In 2009, for the first time, non-Americans registered more US patents than homegrown inventors, with foreigners receiving 50.7 percent of new patent grants. The number of patents awarded to US residents peaked in 2001. The reasons for the shift are made clear by a few obvious facts. As US universities graduated fewer and fewer scientists and engineers, countries such as India and China have graduated more. In addition, American corporations increasingly have shipped research and development overseas in an effort to lower costs. IBM, for nearly two decades the company that produced the nation's highest patent volume, now farms out much of that work to its labs in India. While IBM still owns the rights to any inventions its engineers develop worldwide, the US Patent and Trademark Office registers the related patents as non-resident ones.

And, finally, taxes on innovation are at least partly to blame. The United States was once the world leader in tax credits for research and development; now, we rank seventeenth, according to the Information Technology & Innovation Foundation, as other nations have used tax cuts to spur innovation.

But at least one tinkering expert thinks the shift originates in something more primal. Dean Kamen, one of America's best-known contemporary inventors, whom I spoke with for the purposes of this book, told me, "Tinkering has changed dramatically, but the principle of tinkering—taking the available technology and assembling it to solve problems and create solutions and thereby create real wealth—is not only part of this country, but it is the essence of what made this country viable." Kamen added, "I think we can talk about we're a democracy and about capitalism. But the fact of the matter is for two hundred years we were the envy of the world because we created real wealth."

While most people won't immediately associate wealth creation with tinkering, the two actions are arguably inseparable. "Real wealth is not a zero-sum game, like moving oil here or moving gold there," says Kamen. "There's lots of wealth out there—but every time a new mouth comes out to feed, if you haven't created new wealth, all you've done is reduced the average for the globe, which now has 6.3 billion people."

The economist Paul Romer told an interviewer in 1999, "There is absolutely no reason why we cannot have persistent growth as far into the future as you can imagine." By "growth," Romer meant growth in value, rather than growth in the number of people on earth, or growth in the number of physical objects. "The way you create value is by taking that fixed quantity of mass and rearranging from a form that isn't worth very much into a form that's worth much more," Romer explained in another interview with *Reason* magazine in 2001. The example he gave was turning sand on the beach into semiconductors.

In other words, our society is based on the constant process of re-arranging things and trying to discover some new combination of value. For the past sixty years or so, the United States has been a hotbed of technological innovation. Is that era coming to an end?

Will innovation move somewhere else? What can the nation do to revive that tinkerer spirit?

Fortunately, evidence of new approaches to tinkering is everywhere. Tinkering schools, places where kids are given the tools and freedom to pursue their wildest ideas, are popping up nationwide. So are so-called maker fairs, the modern-day equivalent of craft shows, with the focus on robotics as opposed to macramé. On the economic front, there are newfangled fund-raising engines for bankrolling nascent projects, such as Kickstarter, Y Combinator, and TechStars, which allow tinkerers to get feedback and financing midprocess.

In the following pages, I will explore these trends and others, both in examples drawn from the past as well as interviews I conducted with contemporary tinkerers.

De Tocqueville, in 1835, wrote, "The greatness of America lies not in being more enlightened than any other nation, but rather in her ability to repair her faults." Indeed, our irrepressible optimism and hope in the face of extreme adversity are qualities that have long endeared us to other countries. Or at least they used to, until we stopped radiating these exemplary characteristics a decade or so ago. Suddenly, it didn't matter what the outcome of a solution to a big problem was, as long as someone publicly stated that an effort had been made. The so-called experts had done nearly everything they possibly could, or so it seemed.

Take the blowout of the Deepwater Horizon offshore oil rig in April 2010. After an explosion that killed eleven workers, the ruptured oil well spewed around 60,000 gallons of sweet crude oil into the Gulf of Mexico. That's an amount equivalent to the Exxon *Valdez* spill, dumped into the Gulf every four days. The situation went on this way for weeks. And then months.

A host of potential solutions were implemented in an effort to terminate what was initially believed to be a relatively minor accident.

First came the underwater remote-controlled robots, which attempted and failed to activate the 450-ton blowout preventer, a valve at the wellhead that was supposed to have automatically cut off the leak five thousand feet below if it sensed a sudden change in pressure.

Then came the controlled burning of the oil pooled on the water's surface. On April 28, crews began an in situ burning, a technique in which a five-hundred-foot-long boom is used to move concentrated pockets of oil to a separate area where they are ignited. Other efforts included dropping a variety of domes on the wellhead and attempting to bring oil to the surface for collection. But as the days passed, the oil kept on flowing. The well was finally plugged in late July of that year, with mud and then, finally, cement. By the time BP was able to cap the well, it had already belched 4.9 million barrels of oil, or 205.8 million gallons, into the fragile ecosystem off the coast of Louisiana. And there wasn't a thing the average American could do about it.

Apparently, the experts—the engineers in charge and those employed by the US government—couldn't do much either. The result was the largest manmade disaster in American history. Then, within days of when the cement plug was finally inserted, President Obama announced that two-thirds of the spilled oil had either evaporated or been removed by cleanup crews.

Even the most credulous of oil industry experts had trouble believing the effects of the disaster were reversed so easily. Regardless of the reality, however, the perception of the Gulf oil spill was that American ingenuity had failed in a time of great need. In the American tinkering paradigm, a brilliant individual should have emerged and somehow found a brilliant solution sooner. But that didn't happen. The incident ultimately lacked much-needed closure. Rather than rise to the occasion with an ennobled, enlightened ingenuity, America did its best to cover its tracks and suggest that the problem wasn't even really a problem.

Wising Up about a Smartphone

✧ ✧ ✧

In August 2010, Paul Krugman, the Nobel Prize–winning economist and *New York Times* columnist, published a piece in the *New York Times* titled "America Goes Dark." He described how the United States, "a country that once amazed the world with its visionary investments in transportation, from the Erie Canal to the Interstate Highway System" was now dismantling its infrastructure. "Local governments are breaking up roads they can no longer afford to maintain," Krugman wrote, "and returning them to gravel."

Krugman's main point was that the US government was not investing stimulus funds in the tools needed for our own economic growth. Three decades of antigovernment rhetoric had convinced many Americans that spending taxpayer funds on anything was a waste of taxpayer funds. But government—the US government, specifically—had built this country into an innovative economic powerhouse by investing in "lighted streets, drivable roads and decent schooling for the public as a whole."

I would take Krugman's point one step further and argue that the American government and people helped the country grow both by investing in innovation *and* by committing themselves to the traditional tinkerer spirit. A sophisticated, cutting-edge infrastructure was the perfect crucible for the kind of innovation the United States embodied.

In this book I would like to make a case for the continued importance of the tinkerer in contemporary life and in the role he or she will play in the future of the United States. This is not another book for miserable white-collar workers praising the virtues of manual labor. The point about the devolution of tinkering in American life is not that we have lost a physical connection to the work that we do. It's that the notion that we can fix any problem or achieve any goal that we set for ourselves has deteriorated into a sanitized, corporatized version of what constitutes achievement.

Tinkering as a cultural force once operated well outside American society's mainstream. Tinkerers, even Ben Franklin and Thomas Edison, were sometimes regarded with suspicion or amusement as they made new things out of what existed around them. Indeed, the true ones still do. Tinkering is disruptive; it challenges the status quo. For an individual, or a small group of individuals, to go against conventional wisdom, and thus drastically increase the risk of personal or professional failure, is no small task, even for those imbued with an instinctive American optimism.

In today's corporate world, tinkerers are often found in the engineering profession, partly because engineers have access to the best toys. And while engineering traditionally has been regarded as a respectable profession, in modern America it has been diminished in a culture that venerates business leaders and entrepreneurs. In his 1975 book, *The Mythical Man-Month*, computer scientist Fred Brooks Jr. described how senior corporate managers were often pegged as "too valuable" to devote their time to technical issues, and thus were turned away from contributing to much-needed innovation initiatives. He told how some laboratories, such as Bell Labs, eliminated job titles to overcome this problem: all employees, whether managers or technicians, were referred to as a "member of the technical staff," essentially negating the unique value of those who did the actual figuring of how to make something work. IBM developed two roughly equivalent corporate ladders to address the issue: a managerial and a technical one. Brooks suggested that managers and technical types be trained to be as interchangeable as possible to strengthen the technical know-how of the senior management team.

The implication Brooks makes is that anyone can master the engineering skills required to innovate, and that managers simply need to be schooled in the ways of inventing the future. This notion is antithetical to everything history teaches us about how innovation occurs. Managers and technicians or engineers have intrinsically different value systems and are motivated to peak performance for entirely different reasons. More bluntly, managers crave order and measurable productivity; innovative engineers require unstructured time and an environment that allows for failure as well as success. When corporations blur these differences, they only distract from the indistinct but determinative contributions professional tinkerers provide in a corporate setting.

Corporate America has grown rigid as it has grown larger. Despite the dot-com era's many images of creative whizzes reweaving the very fabric of innovation, it remains extremely difficult for the free-thinking alchemists of today to perform their peculiar strain of magic and thrive while doing it. Google, which has positioned itself as an innovation engine from its earliest days, sought to eradicate this problem by creating its 20 Percent Time program. Under this unique program, Google engineers are expected to spend one day a week, or 20 percent of their time, working on a project that does not necessarily come under their job description. The idea is that tinkering outside of your basic skill set sometimes reaps some surprising and innovative results. But even Google has its limits, apparently: in July 2011, it shut down Google Labs, a platform open to the public that allowed users to comment on the latest projects produced by Google engineers during their 20 percent time.

Modern-day American companies, especially large public companies, simply find it difficult to justify the inevitable overage of resources required to foster truly free-form tinkering. Even if they appreciate it, their investors rarely do.

A good description of how genuine tinkerers are regarded in the modern American workplace can be found in a 2005 management guide by Cornell economist Samuel Bacharach: "Tinkering goals tend to be incremental improvements in the status quo of the organization. The changes a tinkerer makes are first-order changes that do not fundamentally transform the organization. Tinkerers are concerned about changes in specific rules and operations and tend to be risk-averse." Bacharach contrasts that with what he calls the "overhauling approach" favored by big thinkers concerned with "broader goals." Bacharach's portrayal of tinkerers reinforces what has become the ruling image, that of tinkerer as scattershot madman.

This fundamental misreading of the tinkerer's outlook suggests that another way of telling the story of the modern tinkerer is required. This book intends to serve as that alternate history. Throughout these pages, I will explore the work and mindsets of various modern tinkerers. Some are self-selecting, having presented themselves as the contemporary analogues to Franklin and Edison. Others would never think of themselves as anything as grandiose as that. Indeed, these secret tinkerers generally view their work as far from extraordinary. They are simply getting a job done in the best way they know how.

Surface tinkering versus deeply probing tinkering is another dichotomy I set out to contrast. A common complaint among those who worry about the future of innovation in American society is that today's young people aren't motivated to tinker in the way their forbears were. It occurred to me along the way that many of America's best-known tinkerers were not responding to any stimulus beyond their own curiosity. Truly impassioned tinkerers do what they do because it's fun, not because someone is dangling an incentive in front of them. (Gever Tulley, founder of the Tinkering School in San Francisco, California, whose story is told later in the book, knows this and has built his experimental educational programs around it.) As a result, some tinkering is debunked as mere careerism and some is revealed as hidden tinkering, or tinkering in the rough. Surface tinkerers make a big show of their methods, process, and the fabulous end products of their tinkering, whereas deeply probing tinkerers produce innovative thought regardless of the medium, changing the way we think about thinking about things.

The other debate that infuses this book is the relative value of manual tinkering versus digital, or virtual, tinkering. Recent American history is full of examples of tinkerers who have innovated in worlds that exist only on the balance sheets of corporations or in the ether of the computer cloud. Indeed, these new-age tinkerers now outrank traditional tinkerers in both numbers and economic influence.

That is not as worrisome as some observers claim. The canard that "we don't make anything anymore" has the ring of truth, since there is no denying that much manufacturing appears to have migrated to countries where the living wage is much lower than in the United States. But, in fact, more manufacturing still happens on American soil than in any other country on earth. United States manufacturers created around $1.7 trillion in goods in 2009, according to United Nations statistics, outproducing China by more than 40 percent. So why is there a perception that Americans are losing the manufacturing battle?

The answer is simple. The solution is complex. The simple reason for America's continued dominance in manufacturing is that US companies have figured out how to manufacture stuff with fewer workers. Productivity due to innovation has swelled dramatically over the last thirty years. Since the middle of 1979, when manufacturing employment hit its zenith with 19.6 million workers, the US economy has shed around 8 million factory jobs. Meanwhile, American manufacturers have abandoned industries with low profit margins, such as shoes, consumer electronics, and toys, leaving emerging economies such as China and Indonesia to make many of those goods at a fraction of what they would cost here.

American manufacturers now churn out mostly expensive, specialized products that require skilled labor, such as computer chips, fighter jets, medical devices, and industrial equipment. Stateside companies also make anything that requires a quick turnaround time, such as specialized parts for high-tech industrial lathes, which are also made in the United States. Thanks to superior roads, reliable electrical grids and a steady supply of clean water, American businesses excel at producing goods that must be world class.

All of this productivity is, of course, cold comfort for the approximately 14 million Americans who were counted as unemployed in 2011. But the United States arguably rests in a better spot in the global economy than ever before. As long as the nation can continue to produce educated, highly trained workers, there will continue to be a worldwide demand for its goods. In the same way that the United States transitioned from a nineteenth-century agricultural economy to a twentieth-century industrial economy, it will transition again to a high-tech economy in the twenty-first century. For many, the shift will be a painful one, but in the long run it is the one most likely to result in sustainable growth and low unemployment.

Tinkering is not a calling for everyone. But preserving the habitat of the tinkerer is one of the few time-proven ways we as a nation can get back on track. We can't know the future in any way except to know that it will be different than today. Tinkerers acknowledge that in their seemingly haphazard ways. They can't tell you what progress is, but they'll know it when they see it.

Tinkering is a state of mind as much as it is a mode of discovery. The motivations that Americans have had traditionally for creating solutions to the world's problems are as varied as they are vivid. I suspect, as I detail in this book, that what we're really talking about is a crisis of national confidence rather than a systemic failure.

CHAPTER **2**

TINKERING AT THE BIRTH
OF A NATION AND BEYOND

B ENJAMIN FRANKLIN IS OFTEN REMEMBERED as America's first
tinkerer. But the honorific could just as easily have been attached
to George Washington. It is worth examining both men's extrapolitical
activities to help define the scope of tinkering's role in the earliest
years of the United States of America and to better understand how
tinkering came of age in the contemporary era.

Indeed, many of the Founding Fathers were tinkerers of one
kind or another. Thomas Jefferson invented the hillside plow, the
swivel chair, and the macaroni machine. James Madison devised a
walking stick with a built-in microscope to observe organisms on
the ground (unfortunately, it was too short for most men, other than

the five-foot-tall fifth president). And Alexander Hamilton, a fitting forebearer to today's financial tinkerers, established the federal public credit system and the US Mint.

It's hard to say with precision why so many of a small group of political figures and statesmen were also inveterate tinkerers. Some of the reasons are obvious. These were learned, curious men who lived in a time before conveniences such as electric light and time wasters such as television. Perhaps tinkering was a way to exercise the mind, or even to relax it.

The spirit of possibility was also in the air. Not to put too fine a point on it, but these men created a nation out of an idea. In comparison, the notion that objects and institutions could be willed into existence from nothing didn't seem particularly far-fetched.

But over the centuries, Franklin has endured as the prototype for American tinkering. It simply may be because he generated so many inventions and discoveries. As nearly every US student learns, Franklin was the inventor of the lightning rod, the Franklin stove, bifocals, the odometer, and the armonica, an odd musical contraption he designed using a series of glass bowls to create notes based on a man he saw playing melodies on wineglasses in England.

His experiments with electricity became a fulcrum of the industrial age. Franklin embodied tinkererdom in both the traditional and modern senses: He lacked a formal education; he was a dilettante steeped in experimentation but also was appreciative of the fanciful nature of his activities. He had a passion for discovery that seemed to exceed any practical need for the products of his labors, except when it came down to doing business, which he engaged in quite readily.

Walter Isaacson writes that Franklin "had neither the academic training nor the grounding in math to be a great theorist, and his pursuit of what he called his 'scientific amusements' caused some to dismiss him as mere tinkerer." Celebrated as the best-known scientist of his era, Franklin indeed became elevated beyond tinkerer, based on his experiments with electricity alone. But he also was a member of what Isaacson describes as the "upwardly mobile meritocracy," an intelligent social climber who certainly would have been at home in our information-saturated modern society. He was certainly not an engineer: He lacked the purpose-driven focus, never mind the advanced schooling, that defined the profession. But his openness to discovery and his optimistic drive for self-improvement offered something even better to the young nation: a way to remake the world based on one's own interests.

George Washington had a completely different reputation than Franklin, and a different way of viewing himself. Washington was a leader and a war hero—a tall, imposing man with great physical strength—a classic type-A personality seemingly unencumbered by self-reflection or a need for extraneous hobbies.

However, there's another way to look at the first president that casts him as a tinkerer every bit as passionate and creative as Franklin. Washington, both prior to being elected president and after having served, viewed himself primarily as a farmer. But Washington was no ordinary farmer; he was a farmer of the highest intellectual order and innovation. "[Washington] was one of America's first experimental agriculturists," wrote author and educator Paul Leland Haworth, "always alert for better methods, willing to take any amount of pains to find the best fertilizer, the best way to avoid plant disease, the best methods of cultivation, and once declared he had little patience with those content to tread the ruts their fathers trod."

But how would the resourceful general find the better methods? Since there was no agricultural society or agricultural newspaper in the whole country in the late 1700s, he was forced to write to specialists in England for advice, but they were unfamiliar with America's climate and soil conditions. By default, Washington was forced to rely on his own scientific experimentation to improve his farming methods. And so in 1760, he planted a variety of crops including clover, rye, spelt, trefoil, and timothy at Mount Vernon that were heretofore unknown in Virginia agriculture. At the same time he experimented with various fertilizers, including cow dung, sheep dung, marl, and black mold. Meticulously tinkering with different combinations and tracking the results, he decided that sheep dung and black mold were the two most effective. Dissatisfied with the operability of the plows of the era, Washington, in 1760, devised one of his own invention "and found She answerd very well."

That Washington was a man of many public accomplishments is well known, but it is less familiarly acknowledged that innovative farming was a pursuit he maintained through his adult life.

Washington's interests also extended to engineering, though in its early years, America had virtually no one trained to design and build large infrastructure projects. But he did have a vision for extending the country's infrastructure and, after his presidency, pressed Virginia governor Benjamin Harrison to develop a company to help connect Virginia's east coast with the Ohio Country.

Thanks to his status and clout, Washington, who ended his second term as the first United States president in 1797, became president of the newly formed Patowmack Company in 1785, which was founded to improve the Potomac River as a route for commerce. Within the company's charter was a requirement to maintain a navigable channel through the Potomac River of at least one foot deep year round. For nearly forty years, the Potomac had been talked about as valuable transportation route to the West, both from a military standpoint and as economic stimulus.

Washington already had a personal passion for the Potomac River as a conduit into the country's interior, both because he owned western land and because he believed it was a key chance for the young nation to survive and prosper. However, he simply did not have the formal education to make his passion a reality.

Washington tried to hire American civil engineers to undertake the planning, design, and construction of the Potomac Canal. But there were none to be hired. No one in America knew how to build canals. England and France had engineers, but the cost of bringing them to America was prohibitive.

The Patowmack Company occasionally used English engineers already in America as consultants, but the Potomac River was physically very different from most waterways in Europe, limiting the value of their knowledge. Most canals in Europe essentially consisted of man-made underwater steps, or level ditches, that led through a series of locks, or walled pits that raised and lowered the water level. Paths alongside the canals were used to tow boats safely and efficiently through the water passage. The distances were relatively short and the terrain was not too hilly. By contrast, the Potomac was a mountain river, and the distance that needed to be traversed was nearly two hundred miles. The banks were craggy and the vertical drops quite significant. The river was also prone to serious flooding.

Washington and his board of directors ultimately made the engineering decisions for the canal, though Washington took the visionary leap to get the project started by hiring James Rumsey, a quirky tavern owner and builder who knew nothing about building canals, as the company's first technical advisor. Washington previously had hired Rumsey to erect a barn and stable on a property he owned in Bath, Virginia, while staying at Rumsey's nearby inn, called the Sign of the Liberty Pole and Flag. At that time, Rumsey showed Washington a model of a mechanical boat he had invented, which could climb upstream due to a series of poles controlled by a paddlewheel. Washington thought it would be perfect for the canal he was planning.

Washington's approach to creating the canal was pure improvisation. Few people in America had ever seen a canal lock before. He knew he would have to create locks at Great Falls, where the river builds up speed before heading over a series of steep, jagged rocks, and expected he would eventually have to import an engineer from Europe to design them. Meanwhile, he decided he'd just open the channel as best he could. He put Rumsey in charge of clearing rocks from the river bottom. Rumsey, however, soon discovered that the actual-size versions of his mechanical boats didn't work as well as the model. He tried to add a steam propulsion element to his design, but that raised the cost of production dramatically, making it ultimately unfeasible.

Washington and his board of directors' most crucial engineering decision was to opt for sluice navigation, a primitive gate system that diverted water into channels alongside the river, instead of a more advanced lock technology. It would take more than a decade to implement the approach at Great Falls, due to work delays and funding problems. After hiring a series of advisors, including William Weston, an English engineer employed by the Schuylkill and Susquehanna Canal Company of Pennsylvania, the Great Falls section of the canal was finished in 1802, a couple of years after Washington's death, followed by Little Falls, Payne's Falls, and Stubbeville Falls, among others.

Transportation along the river and canals soon became busy during the seasonable high-water periods. Unfortunately, that only amounted to about ten days in the fall and thirty-five days in the spring. Two early American-born engineers, Thomas Moore and Isaac Briggs, later showed that the decision to employ the sluice navigation approach was not only wrong but counterproductive. Sluice navigation made the river more dangerous and difficult to maneuver, due to the unmanageable water levels. Clearing mud and rocks was a constant and arduous chore.

Even worse, the sluice gates required frequent and heavy repairs and wasted excessive amounts of water, a serious issue during the dry season. The lower wooden gates at the Great Falls were particularly susceptible to natural decay; during the summer of 1818, two of the gates gave way and had to be replaced with stone ones. By 1825, many of the gates had deteriorated beyond repair. The Potomac Canal was closed down in 1828 and the Patowmack Company's remaining assets and liabilities were turned over to the newly formed Chesapeake and Ohio Canal Company. The C&O Canal, also known as the "Grand Old Ditch," would run parallel to the Potomac River, connecting the Chesapeake and Ohio Rivers and running from Cumberland, Maryland, to Washington, DC. It operated from 1831 to 1924, though it was made obsolete by 1850, when the Baltimore and Ohio Railroad reached Cumberland.

Benjamin Wright, known as the father of American civil engineering, led the planning and design of the C&O Canal. It was during the execution of his previous project, the Erie Canal, that Wright had stumbled upon entirely different methods of canal construction than those used on the Patowmack Canal. From its use of detailed plans and precise instruments, to the way in which it divided up key projects into individual contracts monitored by a large corps of engineers, the company's approach was unlike any previous one undertaken in the United States. Somehow, between the beginning of the Patowmack Canal and its demise, American civil engineering was born.

The reasons that George Washington is not remembered as a great tinkerer are multitudinous but the biggest of all may be that, unlike Benjamin Franklin, Washington was a *failed* tinkerer. Despite his best efforts to pursue his wildest visions to their logical conclusion, the product of his creativity was not completed during his lifetime. And when it was, it withered and died an ignominious death.

He also had some other accomplishments to fall back on.

✧ ✧ ✧

My point here is that innovators aren't always the individuals who present themselves as such. This is in part because tinkering is an extremely personal and oftentimes solitary endeavor, not conducive to the broad gesture. Furthermore, the fruits of serious tinkering don't always reveal themselves in the short term. It can take years, even decades, for the societal impact of tinkering to be fully realized. Lastly, Americans instinctively favor physical tinkering, the act of creating objects, over virtual tinkering, the act of creating something new that does not result in an immediate material object. On one level, that makes sense—we Americans are a practical, pragmatic people—but it sometimes results in an inability to recognize pure brilliance if isn't right in front of our noses.

This is a situation we can change.

Even those who embody the American tinkerer legend sometimes have had a bigger impact away from the discoveries or inventions most frequently associated with them. Ben Franklin's grand accomplishment as a tinkerer may not have been any of the ones most readily associated with his inveterate puttering, but rather his establishment of the US Post Office in 1775. As publisher of the *Pennsylvania Gazette* in the 1730s, he had publicly clashed with a rival publisher, Andrew Bradford, who printed the *American Weekly Mercury*. Unfortunately for Franklin, Bradford simultaneously served as the postmaster of Philadelphia and exerted the power of his position to prohibit Franklin's *Gazette* from being distributed officially. Franklin was forced to bribe postal carriers to get his newspaper delivered, even after reporting Bradford to the postmaster of the colonies, Colonel Alexander Spotswood.

In 1737, he was able wrest the Philadelphia postmaster gig from Bradford after the latter was called out for his poor bookkeeping practices. Unlike Bradford, Franklin prided himself on delivering competing newspapers; as postmaster for Philadelphia, he delivered Bradford's *Mercury*, as well as the *Gazette* (at least until Bradford failed to pay debts he had accrued while postmaster).

By 1753, he had been named deputy postmaster of the colonies, sharing the job with William Hunter of Virginia. While Franklin took the opportunity to enhance his publishing portfolio and hand out plum jobs to his relatives, he also used the powerful position to make the postal system more efficient. Among his innovations were the first home delivery of mail, a dead letter office, and post office inspection tours focused on improving service. In a year's time, he whittled the time it took to mail a letter from Philadelphia to New York down to only one day.

I hope to underline here that true tinkering is a state of mind, not a set of interests or skills that together somehow form an arrow pointing to the future. Franklin's establishment of the post office had a bureaucratic element to it that may have obscured some of its brilliance. It also was not something that happened overnight. It did, however, require rethinking preexisting elements of American society and reordering them to create something entirely new.

From a time shortly before the formation of the United States through the bulk of the twentieth century, American's character was redefined over and over again by these kinds of disruptive bursts. The country's slow but inexorable progression from an agrarian society to an industrial behemoth was not a simple result of inertia, but rather the result of a series of free-associated ideas that took shape and acquired purpose in the hands and minds of tinkerers, men and sometimes women who saw potential in thinking differently and solving problems the country often didn't even know it had. This is how the country progressed and grew.

In the wake of the second industrial revolution, which spanned from the 1860s to the 1920s, the big problems to be solved no doubt grew more complex. The emergence of electrification, gasoline engines, chemistry, and thermodynamics pretty much insured that most tinkerers from this era onward would need more than a passing interest in these new technologies to make names for themselves in what already had become one of America's best-known exports: the business of solving other people's problems.

Notions of American exceptionalism had hovered around the cocky, young nation from days of manifest destiny. And the ongoing influx of immigrants throughout the original technological age nearly guaranteed the United States' role as ground zero for citizens eager to fix what they didn't like about where they originally came from, especially if there was money to be made.

But by the late 1800s, most major innovations had become science based rather than mechanically based—think cotton gin (1793) versus photographic film (1885). This changed the equation immeasurably. It wasn't as if the average person could have come up with the idea for the telephone or the motor car; this took deep knowledge of physics and chemistry. Over time, this most democratic of countries had, through no fault of its own, erected barriers that deterred the casual handyman from reaching the highest of echelons of fame and fortune. It was one of the ironic by-products of unfettered civilization. Here you were in a land without social classes or inherited power, as close to a meritocracy as the world had ever seen, but it seemed as if everyone you'd ever heard of was smarter and more capable than you.

You would have expected this hard fact to have a chilling effect on a nation brimming with nosy but know-nothing amateurs. After all, why keep trying to come up with something new when you know that in all likelihood someone else has beaten you to the punch? Not because they have some preordained advantage but simply because the free market of ideas is far more thickety and primal than you ever could have imagined.

But, in fact, these unusual circumstances had the exact opposite effect. The final twenty-five years of the 1800s represent the most rapid period of economic growth the world has ever known. It was a time of increased mechanization, furious factory building, the establishment of speedy transportation grids, and enhanced communication networks. Productivity growth during this period went through the roof. And individual prosperity, particularly in America, reached previously unknown heights.

The effect was colossal, igniting what became known as the "American century." Most of the innovations produced by the United States in the second half of the nineteenth century comprise what we know of today as modern life. Suffused by this remarkable change in lifestyle within the course of one or two generations, Americans embraced their new-found primacy and the United States became the dominant economic force in the world,

But at the same time America began to flex its now formidable financial muscle, the enormous impact of its technological innovations seemed to dwindle. Between 1876 and 1900 came the telephone, the refrigerator, the lightbulb, AC electric power, the automobile, aspirin, and the assembly line. After Thomas Edison, however, the output of American scientific tinkerers seemed somewhat diminished. From 1900 to 1925 came air conditioners, toasters, ice cream cones, and traffic lights. With the exception of the airplane, the early twentieth century hatched relatively few gadgets of import.

So what happened?

My personal theory is that the tinkerers went underground. That is, they reacted to the industrial world that had grown up around them by channeling their energies away from the mainstream toward less outwardly identifiable projects. Tinkering became a way of creating systems and organizations as much as a way to create a specific device or machine.

After all, as the world got more technologically complex, so did the problems that needed solving. While the invention of the automobile was an earth-shattering innovation, by the 1920s cars had created a host of new problems: mainly increased traffic and inefficient cross-country transportation. So it's not surprising that one of the main innovators of the era, a man whose tinkering with the way the nation's highways were built, reshaped the way America thought about commerce. And chances are, you've never heard his name.

Born in Leadville, Colorado, in 1891, and raised in Montezuma, Iowa, fifty miles east of Des Moines, Thomas Harris MacDonald witnessed firsthand the frustration of a vibrant farming community limited by the lack of asphalt roads. The same rich soil that made for an abundant harvest also covered the town of Montezuma in mud for nearly four months of the year. "It had the consistency of thick and sticky horse glue," MacDonald's daughter later recalled. "When it rained, you were stuck, your wagons, your feet, you just stayed in your house until it dried. That could be two, three weeks, a month."

As a boy, MacDonald worked at his father's lumber and grain store and watched as business halted as soon as the rain arrived. He later attended the Iowa State College of Agriculture and Mechanic Arts in Ames, one of the many land-grant schools of engineering established in the United States in the late nineteenth century under the Morrill Act. Intrigued by the prospect of finding practical solutions to some of the problems posed by nature, Macdonald was determined to become a civil engineer.

At Iowa State, MacDonald fell under the influence of the school's dean, Anson Marston, a strong proponent of the emerging "good roads" movement. Inspired by Colonel Albert Augustus Pope, a Civil War veteran who sold the country's first "safety bicycle," the good roads movement was founded to protect the legislative right of cyclists.

Think about that: modern highways were conceived of to increase the popularity of bicycles, not cars. It took an awful lot of tinkering to retrofit that idea into the highway system that ultimately helped make the United States both an economic and military superpower.

Learning of his pupil's experiences growing up, Marston encouraged MacDonald to write his senior thesis on the need for highways in Iowa farm country. Shortly after MacDonald's graduation, when the state legislature gave $3,500 to Iowa State College in 1904 to form a committee to study Iowa's highways and how they could be improved to help farmers, Marston appointed MacDonald the committee's chief engineer at a salary of $1,000. At the advent of the automobile age, MacDonald became the first evangelist for highways. In his job to improve Iowa's roads, he discovered rampant fraud in the construction industry that compromised the safety of the state's bridges and culverts. Many would quickly fail, allowing disreputable construction firms to rebuild them again. Traveling the state by horse and by train, the young MacDonald was a standard-bearer for sound construction practices.

A somber man partial to single-breasted dark suits with matching vest and tie, MacDonald exuded authority, though he also was a private man uninterested in cultivating a public image. The short, stocky MacDonald nonetheless became one of the most powerful and influential forces in twentieth-century America, an engineer with a tinkering spirit who recast the nation's roads in a plan of his own device.

In 1919, MacDonald was tapped by the secretary of agriculture to become chief of the federal Bureau of Public Roads in Washington, DC. His success over the next fifty years was determined by this one job offer. It was both a matter of being in the right place at the right time and having the right skills to get the job done correctly. In July 1916, President Woodrow Wilson had authorized the Federal Aid Road Act, which granted $75 million to the federal Bureau of Public Roads. But World War I interfered in the bureau's progress, creating material and labor shortages. Engineers were taken away from state highway agencies and sent to Europe to help the war effort, and constant military traffic from Midwest and East Coast headed for Europe shredded roads that were poorly built to begin with.

When Thomas Harris MacDonald arrived three years later in Washington, the Bureau of Public Roads was in a major rut.

First of all, the bureau had spent a mere $500,000 of the $75 million it had been funded with and had constructed only twelve and a half miles of highway. The federal oversight of the organization, originally considered to be its greatest asset, proved to be its undoing. Federal regulations and imperious federal engineers slowed down construction.

Second, despite a federal mandate, there was no requirement that roads constructed in one state or county link up to those in others. Improved stretches of highway were often stranded in largely unimproved areas. A consensus began building among members of Congress that the Bureau of Public Roads should be eliminated and replaced with a national highway commission. The idea was that local road planning would be traded for the federally controlled construction of three or four roads spanning the whole country.

But MacDonald had little interest in consensus. Despite his deeply conservative nature, MacDonald was a tinkerer at heart, intent on drawing from his fifteen years of experience in Iowa to solve a problem that would determine America's future in a way few understood at the time. There were two things he had come to understand as imperatives for getting things done in highway construction: technical expertise and cooperation.

Informed by the concept of federalism, the evolving partnership of state and federal governments, MacDonald began crafting a revolutionary approach to building modern roads. While political seeming in nature, MacDonald's perspective was actually forged from his years of trial and error as a civil engineer focused on road construction. He learned during those years that maintaining an openness and desire to find common ground among dissenting interests was key to creating a well-operating network of roads. He thought of the highways as a machine that needed to be tinkered with to achieve optimum efficiency. He had experienced in Iowa that local road-building left solely to its own instincts had a tendency toward corruption. And the ineffectual Bureau of Public Roads showed him that a purely national approach could lead to confusion and waste.

Rather than viewing the national highway system as a network of roads, MacDonald viewed it as a network of organizations. Initially, he welcomed any organization that supported his cause. Among the groups he recruited were the American Automobile Association, the Rubber Association of America, the Portland Cement Association, the National Paving Brick Manufacturer's Association, and the American Road Builders Association. In fact, MacDonald welcomed any group that endorsed his cause of constructing roads with federal aid girded by proven engineering and economic means.

When he couldn't find an organization that addressed the issues he cared about, he created one: the Highway Education Board. Remarkably modern seeming in both its mission and its scope, the board was designed by MacDonald to convince Americans of the vital nature of a national highway system. The group distributed fact-laden booklets and films to schools. Its speakers lectured to school assemblies. It even ran essay contests for high school students and awarded engineering college scholarships.

But perhaps his greatest creation was the American Association of State Highway Officials (AASHO), which MacDonald founded in 1914. The name made it sound like just another cog in the bureaucratic machine, and it later turned into one of the most influential lobbying groups in Washington. But its genius wasn't in its ability to lobby members of Congress, though it excelled at advising legislators about highway matters and even assisted them in writing legislation, but rather its role as a nexus for technical expertise. As one of the nation's first technocrats, he built a system of scientific procedures to ensure that roads were being built with the best and most appropriate materials available and in the proper size and location.

The Bureau of Public Roads became a center for research that conducted meticulous studies relating to the best ratio of sand for mixing concrete to the proper pouring conditions and curing times. Subsequently, Congress came to rely on and trust the accuracy and detail of the bureau's reports, which couched its analyses of highway needs and conditions within a rigorously tested body of facts; and state highway departments soon began creating their own research labs, in an effort to apply MacDonald's principles to their own local road conditions.

In collaboration with General John J. Pershing, MacDonald created a chart of roads needed for military defense routes known as the "Pershing map," which became the blueprint for an interstate highway system.

MacDonald spoke of his highway machine as a "complete and economical highway transport service throughout the nation." At an American Association of State Highway Officials annual meeting in 1926, he compared it to what he identified as the only two other "great programs of highway building within recorded history": the Roman Empire under the rules of Julius Caesar and Constantine, and Napoleon's France. The US program was the only one, he pointed out, that had occurred in a democracy.

Thomas Harris MacDonald had the breadth of vision to lay out the fundamentals of the interstate highway system, but not to predict its phenomenal aftereffects. For example, MacDonald was nearly fanatical in his opposition to toll highways, fearing they hindered "freedom of the road."

By the early 1920s, the United States was the world's dominant car culture, with 9 million autos on the road, representing 90 percent of the cars worldwide. The threat was no longer whether there would be enough highways to foster the American economy and the country's military prowess. The threat now was whether there would be enough road capacity. The dramatic increase in road traffic prompted legislators to push for more funding for highway construction.

The burgeoning industries surrounding the manufacturing of automobiles also had an interest in an acceleration of road building. Steel workers, rubber manufacturers, gas station owners, insurance firms, construction companies, oil refineries, and cement plants all had a major stake in the highways of the future. The idea that the need for more highways would be a cause that needed promoting by a team of Washington lobbyists rapidly fell to the wayside. Though, of course, the lobbyists stayed and continued to hammer their agenda far after its path was clean-windshield clear.

By 1936, MacDonald had become an interstate-promoting juggernaut. That year, the federal government provided $225 million for highway building; MacDonald participated in 160 meetings with 85 different members of the House of Representatives and Senate. He also spent ample time before House and Senate Committees making his voice heard on highway legislation.

But MacDonald's most challenging battle would be against the construction of the Pennsylvania Turnpike. Franklin D. Roosevelt had summoned MacDonald the year earlier to reveal his plan for a series of transcontinental interstate toll highways—three east to west and three north to south—that Roosevelt dubbed superhighways. Armed with economic statistics as well as charts, maps, and heaps of construction data, MacDonald made the case for why toll roads didn't make sense from an economic perspective (most drivers couldn't afford them, transcontinental traffic was light).

At first, his mountain of data seemed to have done the trick. The six superhighways remained unfunded. A world in turmoil by 1939 had prompted Roosevelt to focus on the strength of the US military instead of domestic transportation issues. The task of strengthening the ranks of US forces proved to be a sufficient employment engine, at least initially, to delay any transcontinental highway plans. American industry created thousands of jobs to address the new defense needs of the country.

But Roosevelt remained unconvinced by MacDonald's negative assessment. He saw superhighways as a way to spur job growth as well as a means of getting reelected for a third term. Even though defense contracts dominated much of government spending, there was still a small allotment available to build the Pennsylvania Turnpike.

Pennsylvania had long constituted a difficult journey for both travelers and commerce. Its widely variegated topography posed a challenge, including the numerous peaks of the Appalachian range. Back to the era of George Washington and even before, the trip from Philadelphia westward was accomplished at a rate of only forty miles every two days, at best, due to poorly maintained back roads.

In the post–Civil War era, the Pennsylvania Railroad and William Henry Vanderbilt's New York Central Railroad constructed parallel railroad tracks across Pennsylvania. Vanderbilt was trying to retaliate for the Pennsylvania Railroad's construction of parallel tracks to its tracks up the Hudson River. From November 1883 to August 1885, thousands of workers, many of them Italian immigrants, toiled to lay railroad beds and track through rocky western Pennsylvania. J. P. Morgan finally intervened and convinced Vanderbilt to stop building.

But Pennsylvania's traffic needs continued to increase. The only existing main road from east to west was US Route 30, parts of which are still known as the Lincoln Highway. Even into the 1930s, persistent vertical climbs and excessive truck traffic slowed the three-hundred-mile trip from Philadelphia to Pittsburgh to about ten hours on a good day and as many as fourteen hours on a bad one.

In late 1938, the Pennsylvania Turnpike became an inevitability after the Public Works Commission gave its formal approval. While MacDonald continued to disapprove of the plans for a toll road, he was nonetheless cooperative as the plans for the Pennsylvania Turnpike moved forward, opening in October 1940 to rave reviews. In the end, MacDonald was forced to concede that the Pennsylvania Turnpike was a success, both from an engineering and economic perspective. "Every feature of modern road design contributing to a strong, durable roadway and a smooth, uninterrupted flow of traffic has been incorporated in the design," MacDonald later wrote. "The highway represents the best in American practice based on a long experience in road building."

In the next decade, the country built even more toll highways, primarily in the northeast, in anticipation of increased traffic. The Maine Turnpike, the New York Thruway, the New Hampshire Turnpike, and the New Jersey Turnpike became popular motorist routes and generated millions for state treasuries.

The profitability of these interstate highways, which were built not out of need but rather out of enterprise, was a direct refutation of MacDonald's formulation. Still, these new roads were faster and safer than their predecessors, a testament to MacDonald's early push for prioritizing engineering in the road-building process.

As for the fate of toll roads, the Pennsylvania Turnpike Commission had repeatedly promised to eliminate tolls once the bonds used to build the highway were paid off. But that pledge was quietly put to rest, as the needs of the impending war dictated the construction of more highways, which were nicely paid for by the surge in toll revenues.

Still, Thomas Harris MacDonald's vision of an "open source" interstate highway system prevails. While nothing MacDonald imagined directly resulted in a single, tangible object, his tinkering had an immeasurable effect on the way the United States developed during the twentieth century. MacDonald had big ideas, and the time and resources to pursue them. Propelled by his optimism and dilettantism, he was able to shape the nation's roads to a vision he saw in his head. And he did it all as a cog in a larger machine, without the degree of individual recognition we normally associate with tinkerers.

In the decades ahead, however, tinkering would take on a renewed energy as the pursuit of bold individuals, eager to draw attention and prominence to their pursuits.

CHAPTER **3**

CONTEMPORARY TINKERER
FINDS HIS WAY

D EAN KAMEN COMES AS CLOSE AS anyone alive to embodying a
tinkerer in the classic American tradition. At least, anyone with
a Long Island accent. That may sound flip, but it's just my way of
saying that Kamen makes tinkering look easy. In recent years, he also
has developed some strong ideas about how the United States
should fix what he regards as an innovation brain drain and has in-
stituted one of the higher-profile attempts to address the problem.

Born in 1951, in Rockville Centre, New York, Kamen displayed
all the familiar tinkerer traits at an early age. As an adolescent, he
was a dilettante who was naturally good at math but got poor grades
due to his tendency to pursue only topics that were of interest to

him. Meanwhile, he preferred to read books that genuinely interested him such as Isaac Newton's *Principia* and the works of Galileo.

Instead of the usual teenage preoccupations like sports and music, Kamen got caught up in electronics. In the mid to late sixties, one could walk into Radio Shack and find enough interesting electronic parts to figure out how to build something simple like a transistor radio. Kamen instead started tinkering with the latest semiconductors and solid-state supertransistors, particularly ones called thyristors, which can control alternating currents. Kamen realized these neat little devices, frequently used in light dimmers, allowed one to "see" music by synchronizing the sound waves to the lights. Eventually he created a light box, which, when plugged into a stereo system, turned on and off in time with the music. He put on light shows for his friends in his family's basement.

At sixteen, feeling pressure to get a summer job, Kamen followed a lead provided by his uncle, a dentist, who told him he knew the people who worked on the electronics at the Hayden Planetarium at the American Museum of Natural History in New York City. The job he eventually got was working for the man who created slideshows for the museum, among other clients. The job was to build cabinets to house the slide projectors, which typically gave off a lot of extraneous light. The work was pretty menial, and Kamen quit after a few weeks—the job bored him.

But while working at the museum, Kamen got to visit the Hayden Planetarium, built alongside the Museum of Natural History in 1935, and regarded as the most technologically advanced planetarium ever since. But Kamen was surprised by how old-fashioned and cumbersome the planetarium's lighting system seemed. Thanks to his electronics experiments and the light boxes he created, he knew he could improve the synchronization capabilities of the planetarium's lighting rig with SCRs (silicon-controlled rectifiers) and TRIACs (triodes for alternating current). The resulting system would eliminate much of the manual labor then required to put on a show at the planetarium.

Making the most of his access, Kamen barged in to the office of the museum's chairman and tried to sell him on the idea of upgrading the planetarium's lighting system. The chairman, justifiably skeptical of this brash young man, rebuffed him. But Kamen was not to be deterred. Using parts bought at Radio Shack, he designed the sophisticated light show he had imagined, in his basement over the next few weeks. Gaining entrance to the museum with his employee pass, he hooked up his invention to the planetarium's existing light system. The first time he tried it, his circuit board blew up, producing nothing but smoke. Panicked because the summer was nearing its end, Kamen was forced to start from scratch. When he finally got it to work, he invited the chairman to experience it. Angry at first, the chairman was ultimately impressed with Kamen's fully automated system and eventually hired him to install the system at four other museums, including the Chicago Museum of Science. He paid Kamen $2,000 for each system.

Before long Kamen was selling his light contraptions to local rock bands and customizing multiprojector slideshows for other clients— and he had just graduated high school. Kamen would go on to college at Worcester Polytechnic Institute in Massachusetts in 1971, but he had little interest in formal learning, with the exception of courses on physics and engineering. He didn't concern himself with grades and degree requirements. On weekends, he drove back home to manage his light-box business, now known as Independent Prototype. By his sophomore year, in 1972, Kamen was earning around $60,000 a year from his growing business, more than both of his parents' salaries combined.

Kamen's big break came in 1975, after his older brother, Barton, who was in medical school, told him about patients who required twenty-four-hour treatment and had no choice but to come to hospital for treatment. Kamen decided to apply his tinkering abilities to creating a device to administer such treatments at home. He used off-the-shelf parts such as circuit boards, timers, counters, motors, and batteries to build his new contraption. He figured out on his own how to mill his own parts from aluminum.

When William Murphy, the founder of the medical device company Cordis, which had originally hired him to create an audiovisual presentation to promote a pacemaker, visited Kamen in his basement workshop, the young inventor showed Murphy his latest creation, the world's first drug-infusion pump, something he called the AutoSyringe.

Word quickly got out about Kamen's device, and when the *New England Journal of Medicine* wrote about the pump, orders came in from around the world. In 1976, Kamen, at the age of twenty, started a new company called AutoSyringe Inc.

Driven by the need to produce more pumps, Kamen hired a construction crew to lift his parents' house off its foundation while they were on vacation and break through the back of the basement with a bulldozer to expand it into the backyard. Then he hired a crane to lower the equipment he needed into the now sizable workspace, including a Bridgeport milling machine, lathes, and a grinder. Then the house was returned to its foundation, and Kamen's parents were provided with a new backyard patio, thanks to the protrusion of their son's new underground build-out.

Clearly Kamen had more than just the drive to invent. He also had the desire to make a big business out of it.

His next move was to hire away professors and students from Worcester Polytechnic to help him expand his product line. Pretty soon he and his team were inventing infusion pumps for chemotherapy and treating newborn babies, and the world's first portable insulin pump. Kamen kept tinkering, and his team of engineers helped make his ideas a reality. After five years, Kamen finally dropped out of college and began devoting himself full time to his company.

AutoSyringe grew at such a rate that Kamen soon had to relocate its headquarters from his parents' basement to a Long Island industrial space; and shortly after that, to a facility in New Hampshire, a state notable for its cheap real estate and lack of income tax. He convinced twenty of his Long Island employees to move with him.

But while the market for his pumps kept growing, Kamen noticed his own interest waning. By 1982, when AutoSyringe was five years old and had a hundred employees, its visionary founder decided to exit the manufacturing business. At age thirty-one, Kamen sold the company to Baxter Healthcare for a reported $30 million.

Kamen, in his inimitable, unteachable brilliance, had seen a need in the field of health care and filled that need with a solution that, for whatever reason, no one had ever come up with before. Kamen's first insulin pumps were the size of phone books, a drastic reduction from the contraptions they replaced. Today, the Baxter AutoSyringe is as big as a pack of cards, and is the device of choice for continuous delivery of insulin to diabetics.

Shortly after, Kamen founded DEKA Research, the company he still runs in Manchester, New Hampshire. With some of the proceeds from the AutoSyringe sale, he invested in some old brick factory buildings built near the end of the nineteenth century as textile mills. The structures were run down but still solid. Kamen liked the fact that they harked back to a time in American history when tinkering and technology defined the nation.

Kamen went on to invent a quieter, easier to operate, and more portable dialysis machine at the behest of Baxter Healthcare. Dialysis machines at that time were loud, cumbersome, and weighed around 180 pounds. Their technology was controlled by gravity and involved a complex system of valves that had to be connected with tubing. The machines were only available at hospitals and clinics, forcing patients to travel for treatment, sometimes twice a week.

But Kamen didn't try to improve the antiquated machine. Instead, he imagined the infinite possibilities, an approach that made the MBAs on his staff nervous. He liked to joke that if J. P. Morgan had informed the MBAs that he wanted to build a railroad to the West Coast, they would have rejected the notion because of the capital outlay for a railroad that headed toward nowhere.

It was at this point that Kamen solidified his tinkering approach to business. Rather than informing his business with history, he informed it with an unlimited view of the future.

In the case of the dialysis machine, he decided to start from scratch and devised a machine that didn't require a technician to operate, weighed only twenty-two pounds, and was so quiet that patients could sleep while receiving their treatment.

He hired the very best engineers he could find, but, as far as he was concerned, they needed to be more than just smart. He wanted his employees to take big risks and not worry about failing. He called it "frog kissing," referring to the fact that one had to kiss a lot of frogs before finding a prince.

He stripped back the process of building a new kind of dialysis machine to some of the earliest discoveries about physics, Boyle's law and Gay-Lussac's law, and their application to gases. He briefed his engineers on these principles and explained his hunch that they could be applied in a modern context using computer technology, pneumatics, and other new-fangled processes. It took Kamen and a team of engineers five years of tinkering to generate something genuinely innovative that worked. But all that professional tinkering paid off handsomely for Kamen.

Baxter released its HomeChoice dialysis machine in 1993. It wasn't expensive, so patients could buy their own. And it was simple enough to operate that patients could run it themselves. It included a replaceable cassette and easily attachable bags of dialysis fluid. Then all the patient had to do was press the ON button.

The dialysis machine was yet another brilliant product of what had become Kamen's distinct approach to tinkering: taking cheap existing technologies and rearranging them to something amazing and new.

"I don't have to invent *anything*," Kamen liked to say, according to biographer Steve Kemper. "It's out there somewhere if I can just find it and integrate it." He liked to think of himself as a "systems integrator" rather than an inventor. But being the owner of patents, like any other inventor, had its benefits: Kamen's dialysis machine made him a very wealthy man, thanks to the royalties.

Kamen's next gambit was inspired by watching a man in a wheelchair awkwardly maneuvering a sidewalk curb at a mall. He was struck by how many normal situations were prohibitive to wheelchair-bound people. They couldn't go anywhere without sidewalks, they couldn't have eye-level conversations, they couldn't reach high shelves in stores.

He gathered all of the existing wheelchair patents and combed through them for ideas. He was surprised by how many of his own ideas had been tried before. Others had created wheelchairs with legs and arms meant to address the same issues he hoped to conquer, but all of them had one problem or another. None of them inspired Kamen to come up with a new approach.

It took him two years of failed attempts to reinvent the wheelchair before Kamen literally stumbled upon a big new idea that inspired an entirely new approach. It happened one day when he was emerging from the shower and slipped on the wet tile floor. His legs shot forward and, to steady himself, he flailed his arms as a counterbalance to prevent himself from falling. In that moment, he realized he had cracked open his problem. Creating a new kind of wheelchair was all about maintaining equilibrium rather than simply appending walking implements and gadgets that could grab things.

Kamen set his engineers to work on this new approach almost immediately. The prototype that emerged in July 1992 was a homely looking clunky thing, a bundle of bare wires and sharp steel edges held together with Velcro and chunks of foam insulation. But it was what was inside that really mattered: a gyroscopic tilt sensor known as an "inclinometer" first developed for gun turrets on battle cruisers. At $100, it was the appliance's most expensive part. Virtually everything else in the contraption had been bought off the shelf, including two $10 motors made for printers purchased at a nearby discount store.

The chair's core mechanism prevented the chair from tipping over while climbing up stairs or over curbs, even though it only had two sets of wheels. It also could raise the user up to six feet tall and easily travel over sand, gravel, and even up to three inches of water. Kamen dubbed it Fred, after Fred Astaire, due to its graceful stair-climbing abilities. It became better known by its trade name, the iBOT. It eventually sold for around $26,000, with a doctor's prescription, though it was only produced and sold from 2006 to 2009.

He later invented a handful of heart stents, including one used in the body of Vice President Dick Cheney.

But for all of his great successes by his early thirties, the idea for the one that earned Kamen his fame didn't occur to him until he was nearly fifty. The Segway Human Transporter, a relentlessly hyped two-wheel transportation device became a media sensation and then something else. The thing looked like a kick scooter on steroids, and as cool as its technology was (it essentially drives itself), it instantly rendered any adult who uses it forever uncool.

First dubbed Ginger by its creator (as in Ginger Rogers), the Segway built on the balancing technology first developed for the iBot, but with a different purpose. Kamen envisioned the Segway as the primary mode of travel in the postautomobile era. Cloaked in secrecy, the project took more than three agonizing years to develop, during which Kamen's whole company was put in financial jeopardy due to the burdensome development costs. The result was a device that could move forward or backward by the rider merely shifting his or her weight. The effect was a mode of transportation that felt almost like an extension of the human body. Though it also contained motors, as well as a rechargeable battery, the Segway earned its original name with a gracefulness previously unimaginable in mechanized travel.

Again, Kamen had imagined a need where few else saw one, and created something new out of technology that was already around him. In his mind, he was going revolutionize walking. The intellectual power Kamen harnessed to propel the Segway into existence was the ultimate act of American tinkering hubris. But somewhere along the line, the master tinkerer lost control of the Segway's narrative.

In early 2001, a journalist at the media business news website Inside.com published word of a soon to be published book by Steve Kemper about Kamen's latest invention, identified in a leaked book proposal only as "It." The proposal didn't identify exactly what Kamen had created, but it did contain raves from high-profile sources, including one from venture capitalist John Doerr, who claimed that it would be the most important technological development since the Internet; and Steve Jobs's assessment that it would be as significant as the PC and that cities would be designed around it.

Speculation about what "It" was rose to a fever pitch. Between January and August there were dozens of news stories filed about it and countless television reports, all laced with hopelessly hard-to-live-up-to hype. By the time the Segway was revealed to the general public on the ABC morning TV program *Good Morning America* on December 3, 2001, it seemed more like the van in *Stripes* than a world-changing innovation. While the Segway exhibited some undeniably magical properties as a space-age transportation device, among them its ability to balance under any conditions and to move purely from a shifting of the user's weight, its outward appearance suggested little more than a high-tech two-wheeled motor scooter, as dorky as it was cool.

Pretty soon, it was being mocked on the late-night monologues and derided by municipalities around the country as dangerous to pedestrians and, as one cardiologist put it, as "an absurd extension of laziness and slothfulness that will further increase levels of obesity and heart disease in America." Many big American cities banned the use of it on city sidewalks.

By the time the Segway was featured in the 2009 feature comedy *Paul Blart: Mall Cop* starring Kevin James, it was a cultural punchline that needed no further explanation. With just around 50,000 purchased since its 2001 introduction, Kamen sold Segway to HESCO Bastion, a company helmed by British billionaire Jimi Heselden and a group of investors, in January 2010. In a twist that did nothing to help Segway's tarnished public image, Heselden himself died in September 2010 after accidentally driving a Segway off a cliff in West Yorkshire, England.

Today, Kamen chalks up the Segway as another big risk that was worth taking. But these days, his thoughts on tinkering focus more on its future. In a conversation for this book, Kamen imagined a world in which all inhabitants are "enthusiastic and entrepreneurial and risk taking and, in fact, tinker enough to create something that was never there before that is valuable to some large part" of the globe's 6.3 billion residents. Then, he explained, "Every day, instead of having a little less of a fixed commodity, whether it's food or water or gold, every day the world gets a little richer, not a little poorer." Unfortunately, he admitted, that model doesn't exist yet.

But Kamen believes that some version of that model helped forge the United States in its earliest days. "The Revolutionary War ended and you had thirteen bankrupt little colonies that finished a war and had no currency, no credit, lots of debt—like we have today," he says. "But what you had were people who were by definition so free spirited that they formed a country. They took a big idea, an abstraction, and made it real."

By his formulation, it is no surprise that some of the Founding Fathers, these risk-taking political leaders, were also inventors. Europe had its institutions of higher learning, many of which had been around for hundreds of years. But it was the relatively uneducated and unencumbered Americans who invented and innovated to such a degree that they created their own wealth, rather than plundering other countries to acquire it.

"It's a not a coincidence that the United States brought the world all these life-changing, fundamental solutions to problems," Kamen says, "and led the world in creating wealth, and led the world in having that wealth. We didn't steal it from other countries, we didn't go around the world in the zero-sum game of attacking and conquering and taking other people's wealth. That was the way ancient history worked: 'We want your land or your money or your gold.' The U.S. became this country that created wealth."

For decades, America led the world in technological innovation, and melded it effortlessly with what society needed to move forward. "We were the first ones to make cars, and when they became commodities, we started making airplanes," says Kamen. "And when they became commodities we started making computers. And when they became commodities, we started making software. Then we'll do proteomics, we'll do genomics."

Such perspicacity allowed subsequent generations of Americans to make certain presumptions about their own lives. Among those presumptions was one that has become nearly sacrosanct: that each generation will live a better, more prosperous life than the one that preceded it. "Americans sadly think it's a birthright," says Kamen. In a society that continually wants more for each person and where there are increasingly more people, he says, "you can't get there by simply moving the wealth around."

Kamen thinks America's tinkering spirit really lost its way in the area of financial engineering, which contributed to the economic collapse of 2008. "Financial engineering was a way to move wealth around, typically move it from people who had created it to people who were squandering it or using it in some way that was not a catalyst to create more wealth but hoarding it," he said.

Concerned that his country has lost its tinkering mojo, Kamen says the United States needs a new mechanism to reenergize it. "This could be the first generation of Americans that lose that sense of the excitement and the importance and the need to take the risk, to innovate, to try, to fail, to try again," he says. "Every part of our culture was saying there's so much wealth out there; there are easier ways than having to do all that stuff. Let's just use the legal system and the accounting system and just scoop up a lot of this wealth, put it where we want and throw a few crumbs off the table. They leveraged the past, they leveraged their present, they finally leveraged our future. And then it all collapsed."

None of this should have come as a shock, in Kamen's opinion. Indeed, he acts puzzled when describing the impact of the financial collapse on a country ill-prepared to deal with its consequences. "What exactly was the crisis? Oh, you mean the money that was never really there is now gone? I'm not sure that should be a surprise to anybody."

While he acknowledges the world of tinkering has evolved in recent years, Kamen is unwilling to characterize those changes as either beneficial or harmful. "I think it's changed in what we tinker with in some really powerful ways—not good or bad, just powerful.

For instance, you can't tinker by opening up the hood of your car, because most of what's in there is preformed, presealed, very sophisticated. You need custom diagnostic tools, hardware and software tools to figure out what's going on. And you also don't have a group of people who can add that much value or improved that stuff much compared to the people who have now refined it and built it in volume."

Kamen believes the mechanical gizmology of today's world is neither good nor bad. It is simply the end result of decades' worth of successful tinkering having worked its way into the mainstream. And unless one is inordinately nostalgic, it hardly seems rational to pine for a time before the gadgets that help make our lives more productive, meaningful, and enjoyable. On the other hand, there are enough inspired individuals out there to fill up a recent surge in tinkerer workshops sprouting up around the country.

When a piece of technology becomes refined enough that you can no longer tinker with it, people lose the fundamental interest in doing so. A hundred years ago, people tinkered with knitting machines and sewing machines and one-armed bandits. Later on, came the era of Heathkits and transistor radios. But somewhere along the line, the world of tinkering with technology got more complicated: today's computer chips may have half a billion transistors in them and can't be properly examined without a million-dollar piece of hardware.

One thing is for sure: today's tinkerers need more than a little dose of humility and, hopefully, a healthy sense of humor. After all, imagine if the last ten years of your brainpower and entrepreneurial sweat were lampooned (as Kamen's were) in multiple episodes of *The Simpsons*.

But it's not as if tinkering is dead, Kamen assures me. "You could go into a junior biology class today and once a week they have their lab, and we might swab your cheek, put in a couple of reagents and do a little genetic analysis," he offers as an example. His point seems to be that tinkering is not at all what it used to be, but rather what we tinker with has evolved from relatively simple gizmos to unabashedly complex ones. "Do you realize what the average junior high school kid does today in a public high school lab, twenty years ago would have won you the Nobel Prize?"

Kamen's observation seems to be echoed in some trends that have emerged since the economic crisis began in 2008. The prices of high-tech tools and materials have dropped dramatically over the same period, offering engaged young people hands-on opportunities that previously would have been downright unaffordable. The *Wall Street Journal* cited the growing presence of equipment such as modern milling machines, which can craft metal pieces with factorylike precision, in dorm rooms. So-called hackerspaces, fully equipped community workshops where anyone can come tinker with a variety of state-of-the-art gear, are sprouting up all over the country by the hundreds.

SparkFun Electronics, a Boulder, Colorado–based mail-order company started by college student Nate Seidle in 2003, sells all manner of electronic parts and components expressly intended for tinkerers. SparkFun's revenues grew from $6 million in 2008 to $18.4 million in 2010. The Arduino, a low-priced Italian-made circuit board microcontroller designed to operate as the computer core of countless DIY and student-devised electronics projects, has sold more than 120,000 units since its invention in 2005. *Make* magazine, a consumer publication devoted to do-it-yourselfers, started in 2005 with around 22,000 subscribers and increased circulation to 125,000 by 2011. The brightly colored, smartly designed magazine eschews the post-hippie, macramé-weaving vibe of the seventies and instead provides easy-to-follow, step-by-step instructions for projects such as how to build a gigantic bubble generator, how to make your own biodiesel fuel, and how to construct an alarm circuit for a shoulder bag to protect against theft. *Make* has expanded its influence with its popular Maker Faire, which it started in San Mateo, California, and now is held multiple times throughout the year at locations across the country, as well as in Canada and the United Kingdom.

The 1990s brought low-priced personal computers to the forefront, which suddenly allowed independent software developers to compete with large software manufacturers when designing new products. The number of undergraduates who earned mechanical engineering degrees in 2008 rose 27 percent from 2003, according to the American Association of Engineering Societies. During that same stretch, computer-engineering grads declined by 31 percent.

At the same time, spending on research and development has dwindled in the United States, down an average of 2.6 percent per year from 2000 to 2007, based on figures from the National Science Foundation. In the 1980s, that figure was as high as 6 percent.

In most ways, said Kamen, tinkering itself, has not changed. "What's changed is what falls into that category," he said. "That depends on what phase of human technical development you happen to be looking at." At the beginning of the twenty-first century, that tinkering is happening at a very high level, oftentimes in categories, with computer hackers and genomics, that didn't exist twenty years ago.

Kamen is convinced that if the United States does not remain the biggest and best proponent of tinkering, it will lose its position as a global leader. He points to emerging economies like those of China, India, and Thailand as the real long-term threats. As countries increase their economic power they are also quick to realize that their ability to understand and use the tools of technology is directly proportional to their quality of life and standard of living. "That's been true since the discovery of fire," said Kamen.

Things that Americans take for granted such as clean, easily accessible water and electric lights are simple examples of technology we at one point adopted that allowed us as a society to be more productive. Technology and innovation are what took American society from being a population of around 50 percent farmers one hundred years ago to less than 2 percent farmers now, with more food than we could possibly eat.

"The rest of the world is fully aware of why that is," Kamen said. "The great irony to me is that it is only Americans who are clueless, who take for granted that role." The consequence, he argues, is a malaise that has plagued in particular the most recent generations of Americans. The most technologically advanced society on the planet, whose wealth directly stems from that technology, has the one of the lowest percentages of young people studying science and technology in the industrialized world. Interest in these disciplines has declined as other, possibly less valuable diversions have captivated the eyes and ears of the nation.

As an example, Kamen recalls the landing of NASA's MER-A Exploration Rover on Mars on January 3, 2004. The day after it gently touched down on the red planet's surface, seven months after it had been launched from Earth, it began sending back some of the most remarkably vivid color images of Mars ever seen. NASA posted them on its website, which quickly became the most trafficked destination on the Internet in virtually every country around the world. "The world wanted to see this," said Kamen. This American conceived and designed technological miracle had a wow factor that knocked global interest in it off the charts.

According to Kamen, there was only one country where the Mars landing, however, did not achieve top viewership numbers: the United States.

The reason: on that same day, January 3, pop star Britney Spears impulsively jetted off to Las Vegas, where she wed a friend from childhood, Jason Alexander, at the Little White Wedding Chapel. In the United States, entertainment news sites covering Spears's impromptu wedding got the most web traffic in the hours that followed. Never mind that the marriage was annulled only fifty-five hours later.

By pure coincidence, the next morning was the kick-off date for an endeavor related to the educational program Kamen launched to combat what he viewed as correctable problems in America's innovation value structure. Kamen founded FIRST, short for For Inspiration and Recognition of Science and Technology, in 1989, to entice children to become more interested in pursuing studies in science, mathematics, technology, and engineering. Every year, FIRST holds competitions around the country for students, with college scholarships as prizes, often contributed by corporate sponsors. Among those is the FIRST Robotics Competition (FRC), which challenges high school students to build their own robots weighing up to 120 pounds, including batteries and bumpers. Each year's competition has a stated purpose for the robot. In 2011, 2,075 teams participated in FRC competitions in the United States, Canada, and Israel. Kamen often cites FIRST as the invention he is most proud of.

I recently got to see what Kamen described in action at a Mini Maker Faire held in my small suburban hometown in Connecticut. The scientific and technological exhibition featured many cool displays, including a one-man submarine and 3-D printers. But the attraction that by far got the most attention was run by some local high school students involved in a program started by Kamen: the basketball-playing robots. Using some of the gyroscopic balancing technologies developed by Kamen, these young students had created remote-controlled wheeled devices run on cheap motors that could scoop an ordinary basketball off the ground, suck it up through a curved metal ramp, and hook it almost perfectly into the basketball nets mounted on nearby posts. The robots had wheels, but tottered around almost lifelike, as their clearly enthralled creators maneuvered each next shot from a nearby bank of laptop computers.

The crowd, too, was riveted. It was clear they were invigorated by the ingenuity of the young people, in awe of their resourcefulness. The excitement was contagious.

"What will bring this country down, if that's what happens," Kamen said, "is that we no longer, our next generation of kids, are no longer capable of creating the changes, utilizing the state-of-the-art technology to create real wealth."

As smart as I think Kamen is on many topics, and as effective as I believe the FIRST competitions are in getting the young people who participate to think about technology and science, I'm not convinced that he's got a handle on how tinkerers actually influence the American economy. In his alarmist comments, I hear echoes of John Galt's speech from *Atlas Shrugged*. Kamen imagines a society in which only an educated, technocratic elite can keep the American engine running strong, but the reality is obviously much more complex. Even America's great historical innovators, for example, such as Thomas Edison, could be severely ineffective as businessmen, thus on occasion leaving the fruits of their tinkering to be developed by others.

Kamen goes so far as to suggest that America's stagnant unemployment rate is a result not of too few jobs, but of too few educated workers to fill those jobs that are available. "The sad truth may be that it's not that there's a lack of jobs for ten percent of the American potential workforce today, but that there is a lack of competence and capability in the current workforce to fill all the jobs, never mind the really good jobs. If that turns out to be the case, this country better get used to seeing its unemployment rate go up, even if the economy gets better and the stock market gets better and the rich people get richer. And by the way, that's the way the rest of the world used to look until we started this country."

But there is little historical evidence that tinkerers can be trained, though there have been plenty of attempts in the twentieth century to foster formal environments, corporate or otherwise, in which a group of innovators put their heads together and tried to sprout some lucrative ideas. The notion that educating, or rather breeding, a young generation of whiz kids will return the United States to its rightful position as the world's most powerful engine of innovation seems naïve at best and extreme at worst.

Kamen's perspective on high unemployment levels may sound blunt and uninformed, particularly to veterans of the Occupy Wall Street wage inequality protest movement. But some recent studies suggest he may have something of a point. A recent e-book published by two researchers at MIT, Erik Brynjolfsson and Andrew P. McAfee, draws from data they were compiling to write about current strides being made in American innovation. While the authors acknowledge that the ailing economy was the main culprit in the continuing job shortage in the United States, rapidly advancing technology amplified the problem. As work once done by people becomes automated, "many workers, in short, are losing the race against the machine," write Brynjolfsson and McAfee. In the most recent recession, for example, one out of twelve people in sales lost their jobs. During the crisis, many businesses began exploring ways they could use technology to replace the humans they laid off. By the time the recession officially ended in June 2009, corporate spending on equipment and software had grown 26 percent; payrolls remained mostly flat.

As technology has become able to perform tasks once thought to be distinctively human, American workers have not kept up the pace in terms of education and unique skills. Automation was once primarily thought of as the realm of robots in factories, but now it is having an impact on jobs in marketing and sales, as well. Recent innovations such as robot-driven cars and voice-activated personal assistant software (such as the Siri feature built in to the iPhone 4S introduced by Apple in October 2011) suggest that the direction of this trend is unlikely to change soon. The authors of the MIT study agree that the key to improving the jobs scenario is to focus on education and innovation. "In medicine, law, finance, retailing, manufacturing and even scientific discovery, the key to winning the race is not to compete *against* machines but rather to compete *with* machines," they write.

Kamen worries that workers from other countries are not only increasingly able to perform skilled work, once done by Americans, at a lower price, but that these foreign workers are actually more capable than Americans are to do the work they are assigned.

American children are facing more competition globally than any generation before them. "I don't think the fair question is to ask, how are these kids doing compared to their parents?" said Kamen. "The real, more terrifying question is, how are the thirty or forty million kids in this country compared to over a billion people their age in the developing world?"

Kamen, however, claims that he remains an optimist regarding the future of the American tinkering impulse. He laments how the United States used to have the strongest work ethic and the best public education system in the world, available to all its citizens. He also complains that all of our heroes come from the worlds of professional sports and entertainment. "Those are not the source of our wealth," he said angrily. "They are the result of it. They are pastimes." His goal is to make scientists as intriguing as celebrities to teenagers.

Kamen is convinced he can create the Super Bowl of Superior Thinking. To that end, he makes sure that FIRST's competitions are larded with plenty of risks and rewards. The final rounds themselves are set up like major sporting events. "I don't think FIRST exists to address an education problem," he said. "Let's assume that it's not an education problem, it's a culture problem. And it's not a supply problem, it's a demand problem."

CHAPTER 4

EDISON'S FOLLY REINVENTS TINKERING FOR THE MODERN AGE

O DDLY, THOMAS ALVA EDISON'S REPUTATION as the premier inventor of the modern age does battle with his role as the prototypical American tinkerer. Unlike Dean Kamen, who established his credentials as successful entrepreneur, Edison's creative brilliance seemed at times to be nearly eclipsed by his utter ineptness as a businessman. While Edison was responsible for inventing or commercializing an astonishing number of devices that define contemporary society, when it came time to bring them to market, he tended to narrow his focus on some minor aspect of the innovation he had sired into being and lose sight of the bigger picture; that is, the real world in which his inventions would thrive.

It's ironic that Edison failed so much at business since, more than any other American tinkerer, he represents our modern image of the power inherent in combining first-class tinkering with commercial interests to create a dynamic societal force. But we need only look at his singular failure to commercialize the phonograph to understand how Edison's story divided the first century of American tinkering from the second. One might legitimately argue that Edison's failings, and there were many, ultimately served his country better than they served him.

In 1868, at age twenty-one, Edison got a job as a telegraph operator in the main Western Union office in Boston, which was home to one of the oldest and most technically advanced telegraph communities in the nation. Before taking the position, he asked the interviewer whether it would be okay for him to pursue his own projects in his spare time on the job, and was told yes.

So at night he worked the late shift as press-wire operator and, during the day he was free to explore Boston's many telegraph shops and find out what others were doing with the technology. Boston was the Silicon Valley of its day, and there was plenty to discover.

Edison wrote articles for the *Telegrapher*, a trade journal, about the many telegraphic innovations he stumbled upon. The frenetic activity he witnessed also sparked his own ideas. It was around this time that Edison developed his trademark tinkering style of working on multiple projects at the same time. This appealed to the many local investors and corporate officers swarming around the community, and pretty soon, he had funding from a fellow operator, Dewitt C. Roberts, for one of his ideas, a stock-price printer based on telegraph technology.

Among the other devices Edison would develop in this scattershot creative fashion were easier-to-use telegraph transmitters, a fire alarm powered by a telegraph mechanism, and a facsimile telegraph, to transmit pictures and handwriting. Roberts also showed interest and provided funding for another Edison invention, an electric vote recorder, intended to automate the manual voting process then used by state legislatures and Congress. Edison had read about devices of this kind under consideration by the Washington city council and the New York State legislature. His innovation was to incorporate an electrochemical recording technology that was common in automatic telegraphs.

While Edison filed patent applications for everything he invented, the vote tabulator, which included buttons at each legislature member's desk that registered votes on twin dials visible to the chamber's speaker, came through first. Unfortunately, the resultant business had few prospects, since lawmakers were reluctant to speed up a drawn-out process that allowed them to lobby for additional votes.

Another Edison invention, for an improved printing telegraph receiver, earned the backing of a local telegraph company official; as did the fire-alarm telegraph, which lost a contract with the city of Cambridge, Massachusetts, to the largest fire-alarm company in the country. Edison failed to get sufficient funding for his facsimile telegraph, so he simply stopped work.

One might conclude that the reason the young Thomas Edison failed to bring these early inventions to market had to do with the typical difficulties faced by budding entrepreneurs, mainly poor planning, poor luck, and simple business naïveté. But in the case of Edison, that would be the wrong conclusion.

In reality, it was the nature of Edison's tinkering process that was standing in the way of success, not his lack of business acumen. Beholden to his investors, who each had his own strong ideas about the marketability of his inventions, Edison was trying to improve existing technologies rather than letting his mind wander wherever it wanted. Every time he tried to bring out a new version of an already familiar device, done in the name of good business sense, it fizzled due to the extremely competitive environment.

The one exception was his improvement on the printing tele-graph, which was first invented by Edward Calahan. In early 1868, Edison was fortunate that one of the top gold and stock reporting companies was looking to expand into Boston and decided to use his printer for this new outpost. The vote of confidence enabled Edi-son to open his own gold and stock quotation service and resign from his job at Western Union.

From early in his career, Edison had exhibited a "primal and un-varying" need for absolute autonomy in his business endeavors, according to biographer Randall Stross. In July 1869, he left Boston to take a job in New York City, as superintendent of the Gold & Stock Reporting Telegraph Company, located in lower Manhattan. This was the big-league version of the company he had created in Boston. But shortly after that, Edison partnered with Franklin Pope, the man he replaced at Gold & Stock Reporting, to found Pope, Edi-son & Company. New commissions and an influx of cash from their financial backers soon enabled Edison to move to a larger workshop in Newark, New Jersey. By the early 1870s, he had achieved his dreams of running his own workshop, work on projects of his own choosing, and earn a decent living wage doing it.

But even running his own shop, Edison persisted in his practice of working on several projects at the same time. If he got stuck on one project he would "just put it aside and go at something else; and the first thing I know the very idea I wanted will come to me. Then I drop the other and go back to it and work it out." An attorney he knew spoke admiringly of Edison's "remarkable kaleidoscopic brain," which produced countless variations of designs for each of his inven-tions, "most of which are patentable." As for Edison, he frequently wrote in his notebooks at that time, "I do not wish to confine myself to any particular device."

The telegraph that Edison had worked with at Western Union used a series of dots and dashes communicated electronically over wires. By the early 1870s, telegraph companies were on the hunt for any new technology that would allow the transfer of more messages over a telegraph line. One appealing solution was harmonic telegraphy, or acoustic telegraphy, which involved a network of vibrating reeds that allowed the simultaneous transmission of multiple messages. Edison spent much of 1875 trying to perfect his own acoustic transfer system.

Edison was aware of Alexander Graham Bell, just a few hundred miles away in Boston, toiling away on his experimental alternative to acoustic telegraphy. But Edison, like many others, felt the commercial potential of the telephone was limited. Bell's background as a teacher of the deaf kept him focused on certain mechanical elements that aided in the reproduction of speech—Bell's father, also an educator of the deaf, had devised the Visible Speech method of teaching the deaf to talk—even as he also pursued improvements in acoustic telegraphy along the lines of Edison's.

After Bell demonstrated his newly invented telephone at the Centennial Exhibition in Philadelphia, Edison could no longer ignore it. Accounts of Bell speaking over a short distance via his device to British physicist William Thomson and the emperor of Brazil, Dom Pedro, on June 25, 1876, were widely reported. By early July, Edison had started his own telephone-related experiments. A year earlier, he had sought to improve upon an early telephone prototype created by Philipp Reis. His idea was to improve the circuit by transmitting fluctuations in volume and tone by adjusting the current and resistance.

Another event relevant to Edison's evolution as a tinkerer was the moving of his operations and his home in the spring of 1876 to Menlo Park, New Jersey, a farming community twelve miles south of Newark. Menlo Park was relatively rural and isolated in that era, but was conveniently located near the Pennsylvania Railroad line, halfway between Manhattan and Philadelphia.

The Menlo Park compound, some thirty-four acres in total, included the Edison home as well as a two-story laboratory building, one hundred feet long by thirty feet wide, built under the supervision of Edison's brother Samuel. The second floor, which included a balcony overlooking cow pastures, was where Edison and his assistants would conduct most of their experiments; it included a wall of shelves filled with more than twenty-five hundred bottles of chemicals as well as a small work table tucked into a corner that served as Edison's office. The first floor contained a machine shop, complete with a steam engine, as well as a collection of earlier inventions and prototypes, which Edison and his men plundered for spare parts.

It was at Menlo Park that Edison could finally be himself and work in a way that helped him to explore his own natural thought patterns more organically. His tinkering approach already had evolved quite substantially from his early days in the telegraph business, where the business interests of his financial backers had been the primary decider in how he spent his time.

At Menlo Park, Edison laid out his projects in true tinkerer fashion. He typically liked to work on and develop a wide variety of inventions at the same time. Surrounded by a relatively small group of friends and disciples, he would flit from project to project on a daily basis, inspecting work he had assigned to his employees and offering his own ideas for technical improvements. In 1877 alone, those projects included various iterations of his telephone technology, as well as telegraph devices, electric pens, mimeograph machines, sound-measuring instruments, chemical experiments, and even an early incandescent light prototype, which Edison pushed aside after it failed to illuminate for more than a few seconds.

To the outside observer, it would often appear that Edison lacked direction in his experiments. But by this point, the increasingly more confident inventor relied on serendipity as much as experimentation

in arriving at workable, and perhaps more important, marketable solutions. Edison believed strongly in the role chance or accident
played in discovery, something he considered separate from invention
but just as crucial. "Discovery is not invention, and I dislike to see
the two words confounded," he once said. "A discovery is more or
less in the nature of an accident. A man walks along the road intend-
ing to catch a train. On the way his foot kicks against something
and . . . he sees a gold bracelet imbedded in the dust. He has discovered that—certainly not invented it. He did not set out to find a
bracelet, yet the value is just as great."

While Edison was not a scientist in the traditional sense—he was
motivated by commerce as much as scientific discovery—his methods of invention incorporated some of the same approaches as those
of scientists who typically worked on their theories for many years
before achieving a breakthrough moment.

Still, Edison felt the influence of external factors, which had the
habit of sidetracking his preferred approach to tinkering.

Pressured by the public attention on Bell's telephone, Edison was
forced to reconsider his opinion of what he previously termed
"merely scientific toys." Not the ideal person to be working on a telephone, since over time he had become nearly deaf, he nonetheless
soldiered on in his quest. By 1877, Edison was actively working on
what he perceived as the weakest element of Bell's telephone, its
transmitter. In Bell's transmitter, a magnet was vibrated by sound
waves and generated a variable current that, after traveling through
the line, was then turned back into sound waves at the receiver.

While hardly passionate about his telephone, Edison was driven
by his competitive nature to best Bell's invention. In addition to crit-
icizing Bell's transmitter, he ran down its receiver in speaking with
colleagues, pointing out that it could not be used on longer lines
because of poorly modulated current resistance.

Edison cycled through a series of telephone-related experiments over the next couple of years. Many touched on previous experiments he had done with carbon, an element with conductive properties uniquely suited to solving the current resistance problems Edison hoped to redress. In October 1876, he used a stick of Arkansas oilstone coated with graphite, carbon in its softest form, as a resistance medium in place of Bell's magnet. The design he finally settled on, and for which he ultimately received a patent, was a carbon transmitter made of a parchment diaphragm with a tinfoil face that pressed up against a disk of hard rubber coated with plumbago graphite to complete the circuit. This device provided the pliability and sensitivity needed to respond differently to high and low notes, and therefore more accurately reproduce the varying modulations of the human voice.

In late 1877, Western Union officially identified the telephone as a real threat to its telegraph business, after Bell's company began offering and selling private telephone lines to businesses. It quickly started up a subsidiary called the American Speaking Telephone Company, with $300,000 in assets, and snatched up all the related patents it could manage, including Edison's loud and articulated carbon transmitter. Edison's transmitter was so much better than Bell's that the Bell Company would have been put out of business if it had not stumbled upon an unknown inventor named Emile Berliner, who had devised his own telephone transmitter employing scientific principles similar to those developed by Edison.

As a result of a yearlong lawsuit filed by the Bell Telephone Company against Western Union in September 1878 over what it claimed was Western Union's pirating of Bell's telephone receiver technology, the leading telegraph company agreed to withdraw from the fledgling telephone industry entirely. In exchange for the rental of telephones on its existing lines, Western Union received a 20 percent royalty, amounting to $3.5 million, from Bell Telephone. While the deal seemed like a smart one at the time, it gave Bell full control of America's telephone industry.

Edison himself made his own dubious deal for his telephone technology. A few months prior to the Bell settlement, he brokered his own agreement with Western Union for his carbon transmitter, for $100,000, a handsome sum for the young inventor but a remarkably small sliver of the ultimate profits. Oddly, Edison made a special request that his fee be paid in seventeen annual increments of $6,000—he feared he would spend a lump payment too quickly—since the interest on $100,000 in savings likely would have paid that much.

Meanwhile, Edison set out on the task to develop a better receiver for his telephone. Starting in March 1877, he performed a series of experiments in search of an alternate receiver design, many of which used what Edison called the electro-motograph, a cylinder of chalk moistened with caustic soda and turned by a crank. An electromagnetic arm was held with tension to the cylinder by a spring, and when electricity was applied to the arm, it would vibrate and create a pattern on the chalk cylinder. By July, he had settled on one with a speaker composed of a diaphragm with an embossing point held against a sheet of paraffin paper, which he later exchanged for ridged tape not unlike that used for a stock ticker. On the night of July 18, 1877, after a midnight dinner with his workers, who were accustomed to working long hours, he stumbled upon a variation to his original idea while playing with some of the diaphragms they had constructed: Edison realized he could easily record sound and play it back later.

His first thought was that he had on his hands a business machine that far exceeded the usefulness of Bell's infernal telephone. At least initially, telephones were employed in the same way as telegraph transmitters: a company employee would transmit the message verbally, instead of in Morse code, and it would be written down on the receiving end. Edison realized that a more useful device would be able to record the spoken message automatically. In his own account, Edison recalled how he subsequently devised a toy that used the sound vibrations of a telephone diaphragm to power a pulley that made a paper man saw wood. "Hence, if one shouted: 'Mary had a little lamb,' etc., the paper man would start sawing," he explained.

But while Edison's receiver did in fact reproduce sound beautifully, it was impractical as a telephone. Although it was able to receive speech well enough in laboratory conditions, it did little to further articulate the human voice over long stretches of phone line.

So he returned to his idea of an embossing telegraph repeater. In June 1877, he had made notes proposing the use of "thin copper or other metallic foil" rather than the ridged tape he had tried previously in an effort to sharpen and raise the volume of the human-sounding mutterings that emerged from the device upon playback. It wasn't until November, however, that Edison had a detailed sketch of what he thought to be one of his most commercial projects to date: he would produce a telephone repeater, along the lines of the telegraph repeater, that would be useful for recording and reproducing sounds coming through Bell's telephone.

Edison had no idea he was actually inventing something completely different.

The sketch of what he had in mind was a contraption that used "a cylinder provided with grooves around the surface" and wrapped with tinfoil, "which easily received and recorded the movements of the diaphragm." A hand crank would be attached to the cylinder to rotate it. Then he assigned the job what he called a "piece-work price" of $18. This meant one of his company's workmen would get the job as a challenge. If the workman failed to produce the machine properly, he would get his regular pay; but if he succeeded in building a working model, he would get the $18 also. It was an easy bet for Edison, who was not optimistic that what he had designed would work.

John Kruesi, the workman assigned to the project, apparently had no idea what the machine he was building would be used for when he started the job. When it was nearly done, he asked Edison what it was. When Edison told his employee that he was "going to record talking, and then have the machine talk back," Kruesi thought it "absurd."

But when the device was completed—with two diaphragms, each attached to a stylus and mounted on tubes on opposite ends of the cylinder—the effect was nothing short of astonishing. On December 6, 1877, Edison shouted the first few lines of "Mary Had a Little Lamb" into one of the diaphragms while he turned the hand crank. Then he turned the crank backward to where it had begun, removed the first diaphragm tube from the tinfoil, and replaced it with the other. Once again, Edison turned the hand crank forward. His voice played back almost perfectly. "I was never so taken aback in my life," he said. "Everybody was astonished."

On the morning of December 7, Edison and his business partner and associate Charles Batchelor hauled the new phonograph from his Menlo Park workshop to New York City and the offices of *Scientific American*, where he proceeded to demonstrate it for editor Alfred Beach and the other employees of the esteemed publication; according to Edison, so many people gathered around Beach's desk that the floorboards were at risk of collapsing. He set up his phonograph and, as he had done in his workshop, he spoke into it while turning the crank, then reversed the cylinder back to its starting point, switched diaphragm arms, and played the result. The gathered crowd could hardly believe its ears. Many believed Edison's demonstration was a parlor trick, the act of a skilled ventriloquist. The morning papers reported the event in tones of near disbelief. Two days later Edison filed a patent application for his "phonograph or speaking machine."

The resulting *Scientific American* article, entitled "The Talking Phonograph," published in the issue of December 22, 1877, brought Edison something he had neither sought after nor anticipated: fame. Reporters from major newspapers in the region descended on the Menlo Park studio, and were politely admitted by the inventor.

The fact that Edison, who had a plainspoken, down-to-earth manner about him, good-naturedly answered questions posed by visitors no matter how silly or ill-informed, made him even more of a folk hero. He was admired by Americans who believed in the power of hard work, though those who were skeptical of his achievements hinted he had ties to the occult. Branded by some as a "wizard," he became known as the "Wizard of Menlo Park." Cognizant of the many potential uses for his phonograph, Edison liked to say that the phonograph would grow up to support him in his old age. In an article published in the *North American Review* in June 1878, called "The Phonograph and Its Future," he detailed some of them, many of which would become reality: dictation without a stenographer; phonographic books for the blind; listening to music; recording the voices of family members as a keepsake; clocks that announced the time; the preservation of the words of great men; the recording of teachers' instructions for students to refer to at a later date; and the making of permanent records of telephone transmissions. But for all of his sudden fame, Edison again met failure when it came to turning his invention into a business.

Unfortunately, Edison's original tinfoil phonograph was far from user-friendly. It required superior manual coordination to operate. The tinfoil had to be wrapped around the cylinder at just the right level of tension, so that the stylus would leave a deep enough impression but not rip through the foil. Then the operator or a companion had to yell a message for, at the most, ten seconds, while cranking the feed screw at a consistent pace. To hear the recording, a playback stylus was placed at the beginning of the groove that had been made with the first stylus and the operator had to crank the machine at the same speed as it was originally recorded to get an accurate rendition of the original performance.

Furthermore, each tinfoil recording could only be played a few times before it wore out.

Edison had little or no understanding initially about the phonograph's potential as a home entertainment device for adults. He had poor hearing and was not a lover of music. He envisioned its primary commercial use as a dictation machine for business. But when a group of venture capitalists—which included newspaper man Uriah C. Painter and Alexander Bell's father-in-law, Gardner G. Hubbard—approached Edison in January 1878, he agreed to the establishment of the Edison Speaking Phonograph Company, whose primary initial purpose was to distribute a line of cheap toys including dolls, trains, and birds, which would all generate sound via a hidden phonograph. Edison farmed out the manufacturing and sales of these cheap machines, with the intention of funding future inventions with the royalty checks.

The investors provided Edison with an initial payment of $10,000 to refine his invention for the marketplace. Once it was on sale, he would receive a 20 percent royalty on net proceeds. The Edison Speaking Phonograph Company was well aware that the other commercial uses for its product had yet to be established, so it hired the Redpath Lyceum Bureau of Boston, a well-known lecture service, to arrange a tour of exhibitions of the phonograph around the country. For the next year, Redpath shuttled five hundred phonographs to entertainment halls and amusement centers, where visitors were charged a small admission fee to watch barkers demonstrate the machines by playing short tunes or comedians' jokes from tinfoil records that lasted about a minute and a half. At first, crowds swarmed to these exhibitions; Edison's royalty from one such event in Boston totaled $1,800. But after a number of months, the excitement died down, and the novelty of the clever phonograph waned.

Meanwhile, Edison continued to tinker with his invention. By linking the cylinder to a steam engine, he was able to improve its fidelity and volume significantly. Of course, the average consumer was unlikely to have a steam engine at home. So Edison experimented with powering the cylinder with a crankable clockwork mechanism. The clock mechanism solved the consistent revolution problem, but the user still had to crank it 120 turns per minute to achieve a decent playback. Through a process of trial and error he also discovered that using copper sheets instead of tinfoil helped increase the volume of the records.

Edison and his underlings continued to tinker with the machine, intent upon improving it for what Edison remained convinced was its future use, as a dictation machine. Realizing that the shouting required to make a good recording was not practical for an office setting, he sought for other ways to improve playback volume. One idea involved a valve that issued steam or compressed air to resonate a large diaphragm that made the sound of the recording much louder.

But the investors in the Edison Speaking Phonograph Company, wishing to see a return on their money, began pressuring him to get a commercial phonograph ready for the business market. To placate them, he agreed to produce five hundred smaller, cheaper models that could record no more than forty words. The smaller model, to be released in April 1878, would be nothing more than a novelty, intended simply to demonstrate how the phonograph worked, until the bigger model was ready.

It was just before the release of the small model that Edison agreed to an interview with a reporter from the *New York Sun* named Amos Cummings. Cummings had requested the interview at least a month earlier, clearly interested in exploiting Edison's growing celebrity, and the inventor promised he would grant it. After putting it off repeatedly, Edison finally relented and allowed Cummings to interview him at Menlo Park. The resulting profile, which ran on February 22, 1878, identified Edison as "A Man of Thirty One Revolutionizing the Whole World." And it would change his life. The story described Edison in nearly mythic terms. The general public's understanding of technology was limited when it came to something as complex as Edison's phonograph. As a consequence, he was portrayed as someone akin to a Broadway personality. Edison may have been partly to blame, eager as he was to please a now adoring public.

But Edison's newfound fame did little to help his phonograph business's prospects. All of the free publicity and excitement surrounding the phonograph seemed to guarantee its commercial success. But the Edison Speaking Phonograph Company was ill-prepared for the wildfire interest in its yet unreleased product.

There were no serious competitors at the time for his phonograph. He had all the resources he needed to develop it, both financial and material, and an admiring public awaiting his final product. But even with all these advantages, Edison somehow could not deliver on his promise.

The smaller versions of the phonograph were not particularly reliable or useful, and did not produce the sales Edison's investors had anticipated. In May 1878, the company sold forty-six of the exhibition model; in July, only three. September's sales surged to sixteen units, earning Edison $461 in royalties. Meanwhile, a staff of five labored in Edison's lab, rushing to develop the full-size model the inventor had promised. But the men struggled to come up with one that was suitably reliable and affordable.

Edison continued to investigate various improvements to his phonograph over the next few months, including an idea for a wax record shaped like a plate. But new discoveries created new problems: a stylus on a flat disk distorted the performance as it spiraled closer to the center of the circle; then there was the matter of how to produce affordable copies of records. The more he learned, the more Edison came to believe that his dream of his phonograph as the ultimate business machine was futile.

There were plenty of distractions that added to the difficulty of the task. After all, the invention of the phonograph had been nothing more than a detour from Edison's work in the telegraphic industry. His work in Menlo Park continued as it had before, with his scattershot approach to tinkering becoming ever more frenetic. By the middle of 1878, Edison switched his focus to a prototype for a commercial hearing aid. From there, he delved into work on a microphone he designed to replace a medical stethoscope. Neither of these inventions ever came to market. At the end of the year, Edison's royalty payments for the phonograph totaled $1,031.91, mostly from exhibitions. He also had a new prime interest: electric light.

As biographer Matthew Josephson observed of the inventor's failure to capitalize on his telegraph innovation, "What Edison did not then realize, except dimly, was that the decision as to the commercial acceptance or refusal of inventions, and much of the control of industrial technology, turned not upon the question of merit or usefulness, but upon the outcome of intermittent wars or peace negotiations between the rival 'barons' in the railroad and telegraph fields, such as the Goulds and Vanderbilts."

Indeed it was an offer of $50,000 to develop a practical electrical light from the industrialist William Vanderbilt and his friends in late 1878 that rather firmly shifted Edison's focus away from the phonograph.

Edison had attempted some experiments with electric light a few years earlier in Newark, he claimed, over his frustrations with the local gas utility, which had threatened to remove their meter and cut off his gas supply when he had trouble paying his bills. But it was a visit he took in September 1878 to the workshop of William Wallace, the owner of the famed Wallace & Sons brass and copper factory in Ansonia, Connecticut, that reignited his interest in incandescent light. Wallace and inventor Moses Farmer had developed a powerful electromagnetic generator that could illuminate eight electric arc lamps, an early, less sustainable form of electric light.

The visit inspired Edison to imagine something much larger than a single electric light source. Though he had yet to invent the incandescent lightbulb, he quickly envisioned a network of energy sources that would allow the introduction of electric lighting into private homes, much in the same way that Thomas Harris MacDonald would later conceive of the interstate highway system.

In the process of broadening his outlook to envision a radically changed modern infrastructure, Edison also reinvented the process of tinkering for the contemporary era. Unlike the phonograph, the lightbulb was not invented by Edison alone in his lab but rather through a series of inventions generated by one of the first true corporate research organizations.

Up to that point, Edison, though more ingenious than most inventors, had operated in a fairly traditional way. He employed two or three trusted assistants and a couple of skilled machinists who together labored make the ideas in Edison's head a reality. But in the year that followed, he began adding to his staff some different kinds of employees to aid in his experiments with electric light. These new employees were not merely competent lackeys who followed their boss's instructions to the letter, but rather skilled professionals who brought their own ideas and methods of experimentation to the workshop. Among the new staffers were chemists, glassblowers, and a mathematician with graduate training in physics. And at the helm was Edison as director of research.

While Edison continued to tinker in some of the same ways he always had, he understood that creating an electric lighting infrastructure would require more brainpower and coordination than one man, no matter how brilliant, could muster.

It would be another ten years before Thomas Edison would have another shot at marketing a version of his phonograph. In the years between 1878 and 1887, while Edison had focused on other inventions, the phonograph had evolved beyond his original tinfoil cylinder model. Edison had waited too long to file the patents for his original tinfoil machine, and this allowed others to further develop the technology without paying him royalties. Among them was none other than Alexander Graham Bell, whose Volta Laboratory developed a wax cylinder technology that improved the accuracy of the sound reproduction.

In 1887, the American Graphophone Company launched a machine that used durable wax-coated cylinders instead of fragile tinfoil. The fact that one of its three inventors was Bell, who had beat Edison to market with his telephone years earlier, prompted the always competitive inventor to return to the invention that had been his calling card and, once again, begin tinkering with it.

In March 1888, American Graphophone approached Edison, offering to merge his phonograph company with theirs and giving him full control of the graphophone and any improvements to it he saw fit. The company's ploy was clear, however, since it dubbed Edison the true inventor of the phonograph: it wanted to use Edison's name on the graphophone as a way to boost sales. But Edison would have none of it, writing to an associate, "Under no circumstances will I have anything to do with Graham Bell [or] with his phonograph pronounced backward."

Edison hastily constructed his own wax cylinder phonograph with the help of his development team and set out to convince investors to finance its manufacturing and distribution. But an early demonstration in the spring of 1888 failed after an Edison assistant mistakenly replaced the original diaphragm with one that he believed to be an improvement. He also exchanged the recording stylus for a more refined one. Unfortunately, he forgot to change the reproducing stylus, which was broader and blunter. The effect was that the reproducing stylus was too wide to enter the groove created by the recording stylus and as a result produced nothing more than a extended hissing sound.

Edison would never again have the opportunity to start his own phonograph company.

The owner of the North American Phonograph Company, Jesse Lippincott, rescued Edison's newfangled phonograph by agreeing to distribute it; Lippincott also distributed the Graphophone. Edison loathed the arrangement but had run out of choices. He did, however, insist on retaining manufacturing rights.

Unfortunately, the arrangement only ended in financial trouble for Edison. First, many of the phonographs manufactured in a trial run proved to be defective, forcing them to be recalled. Then Edison discovered that Lippincott had made a deal with one of his employees, Ezra Gilliland, to purchase Gilliland's agency contract in exchange for $250,000 in stock in a new company he was establishing to sell both the phonograph and the graphophone. Gilliland split the stock with Edison's attorney. After the deal was done, Lippincott mentioned the deal to Edison, unaware that he had not been told of the arrangement.

Edison sued Gilliland and his attorney for fraud but the defendants were able to prove that they had full permission to act as Edison's agents and therefore had simply perpetuated a breach of ethics rather than anything legally actionable. And having severed all ties with Gilliland, he also in effect severed ties with his last hope at making a fortune off the phonograph. At the time of the incident, Gilliland was preparing to incorporate a company designed to market the phonograph as an entertainment device.

By 1892, it was already clear that the phonograph would succeed in the entertainment field rather than the business world, as it became a popular attraction at nickelodeons, where a placing coin in a slot would produce music or comedy routines.

The relevance of Edison's failure to successfully market his phonograph and his inability to significantly profit from it is that his laserlike focus on the pure act of innovation was by no means a guarantee, *or even a definable asset*, in his quest to thrust his invention into the culture at large. No matter his technical brilliance or his rapidly ascending fame: the maverick genius was simply making things up as he went along and figuring out how they might make sense as a business after the fact.

And that is just as it should have been. After all, that's how most tinkerers come up with something genuinely new. How can you put a value on curiosity? How can you codify a willingness to experiment? And what toll does a regimented system take on the free flow of ideas?

Edison's charismatic creativity, his unschooled smarts, and his stubborn personality all seemed to conspire against the notion that his tinkering, or any tinkering for that matter, could be systematized and mass-produced in the style of his good friend Henry Ford.

And yet Edison's Menlo Park lab became an innovation in its own right as the world's first corporate brain trust. While Edison himself wasn't always able to capitalize on the merits of his unique research and development operation, its mere existence had a major impact on American corporations going forward.

In the shadow of the mushroom cloud that launched America's nuclear era grew the logical next step to Edison's notion of the tinkering conglomerate. On July 16, 1945, the group of scientists who comprised the Manhattan Project watched with horror and glee as the bomb they had built lit up the skies above Alamogordo, New Mexico. By October of that same year, the United States had dropped two more atomic bombs on the Japanese cities of Hiroshima and Nagasaki, horribly maiming and killing countless civilians.

The RAND Corporation, perhaps the best-known think tank ever assembled, was chartered on March 1, 1946. Gaining its name from a conjunction of research and development, its initial purpose was secretive, militaristic, and analytical by design. Established by five-star air force general Henry Harley "Hap" Arnold and former air force test pilot Franklin R. Collbohm, RAND was designed to harness top military minds in the post–World War II era to produce new weapons and strategies to protect the country's interests internationally.

But it was the pursuit of peace that produced the RAND Corporation's most innovative approach to tinkering. In the early throes of the Cold War in 1950, RAND analysts set about to codify human behavior, reducing it to a series of mathematical formulas and equations. Their purpose was to transform warfare from a series of blind attacks into a sophisticated portfolio of tactics and strategies, designed to minimize the destruction of human capital (particularly American human capital).

Game theory was the start. Developed by the Hungarian mathematician John von Neumann, game theory proposed to apply mathematical probability puzzles to human behavior. Neumann's primary assumption, which traced its roots back to an eighteenth-century card game, was that players in a game were rational and therefore predisposed to finding a solution, or a rational outcome, to any problem.

RAND was particularly drawn to game theory and even hired von Neumann to apply his "zero-sum game" principles to the set of problems it had committed itself to solve. The classic game favored by RAND is known as the "prisoner's dilemma."

Imagine two men are arrested for a crime, such as the theft of a precious diamond. The police separate the men, preventing them from communicating with each other. They tell each man that if he says where the diamond is stashed, he will do only six months in jail. The prisoner who refuses to confess will get ten years in prison. If both of the prisoners confess, they each serve a two-year sentence. Of course, if neither man confesses, and the diamond is not found, then they both go free.

The prisoner's dilemma became a stand-in for the arms race that plagued the United States, and the RAND's researchers became obsessed with what it would take for one prisoner to confess—or, as they called it, defect. Unfortunately, when they queried test subjects regarding the prisoner's dilemma, answers tended to reveal the political outlook of the person being asked rather than predict the likely outcome of the game.

Liberals had more faith in their fellow human beings, and were thus more likely to envision a level of trust between the two prisoners. Conservatives more often viewed themselves as defectors, preferring to focus on self-interest and self-reliance as guiding human instincts.

Alas, by the mid-1950s, RAND had largely abandoned game theory as a guiding tenet of national security policy, having decided that there was no one correct solution to the prisoner's dilemma. After Stalin's death, the Soviet Union became substantially less opaque, due to Khrushchev's efforts to communicate more openly with the West; trying to forecast Soviet military policy was no longer the dark art it had once been thought of as.

Looking for a new focus, RAND turned in the mid-1940s to an approach to defense strategy that strikes me as one of the first post–World War II examples of tinkering. In the same way that Thomas Edison assembled a team of assistants and engineers to help envision the world not as a series of problems that needed solving but rather as a canvas on which a master tinkerer could project his vision of the future.

The term used by RAND for its tinkering procedure was "systems analysis," coined by a RAND engineer named Ed Paxson in 1947.

Paxson, formerly a scientific advisor to the US Armed Forces, had long admired game theory, and wanted to apply its ideas directly to the process of war. Inspired in part by what was known during World War II as operations research (OR), Paxson set out to take America's defense policy beyond the statistics and hard data that, in his view, weighed it down.

In operations research, the goals were to figure out how much damage could be inflicted with the resources available to the United States and how efficiently those military plans could be executed to minimize losses. At what altitude should fighter planes fly? How many would be required to achieve a certain goal? Mathematics played a key role in achieving these data-based goals.

Systems analysis was different from game theory, and distinctly American in its outlook. Instead of relying on existing data to craft solutions to the nation's current defense goals, systems analysis first reflected upon what the nation hoped to achieve in the future, and how it might craft solutions to those problems. As Alex Abella explains it, in his well-reported book on the RAND Corporation, *Soldiers of Reason*, "Systems analysis changed the questions and asked: How many enemy factories do we *want* to destroy? What kind of factories are we talking about and how are they defended? To accomplish our objective, what is the best route? With what kind of plane? What kind of payload?"

While operations research focused on finding new ways to improve and streamline existing systems, systems analysis took today's existing knowledge and tinkered with it, until it created the potential problems of tomorrow, as well as an assortment of systems to solve those problems, systems that had yet to be devised.

Operations research amounted to doing the best with whatever one had. Systems analysis insisted on expanding the palette of options, even if those options didn't exist yet. In essence, it was a license to dream. Not necessarily to dream only of the destruction of

America's enemies, but rather to dream of a democratic nation that could bend the will of nature to embrace its own desires.

If that notion sounds a trifle grandiose, the reality was substantially more down to earth. Indeed, in classic RANDian fashion, the dreaming process itself was steeped in mathematics and methodology. Imagined projects were categorized and measured and cost-analyzed as if they were real. If a new fighter plane were part of the equation, systems analysis would determine how fast it would go, how much it would cost to build, how far it would fly, and how much fuel it would use.

Prompted by the detonation of an atomic bomb exploded by the Soviets in 1949, the US Air Force sought to fashion a preemptive attack on the Communist powerhouse. It assigned Ed Paxson to create a suitable bomber for the plan.

Paxson's approach was both creative and disheartening, though perhaps not surprising given the grim goal of the task. Set on developing a "science of war" out of whole cloth, he began by compiling all manner of detail about the aerial bombing capabilities of both sides in the projected conflict. The mountain of resulting data was so large that RAND developed its own early computer to manage the results.

To top it off, Paxson created a combat simulator in RAND's basement to allow pilots from the air force and navy to practice their craft against films of real war footage. And Paxson's resulting report, "Comparison of Airplane Systems for Strategic Bombing," produced in 1950, ultimately betrayed the tinkering roots of systems analysis.

Make no mistake: RAND's approach to systems analysis had a fatal flaw and, as a result, took American tinkering down a dark hole. RAND's so-called rational approach to national defense helped shape the United States' policy of counterforce in the 1950s (stockpiling nuclear warheads) and its disastrous response to the Vietnam conflict in the 1960s.

And while RAND liked to paint its peace-keeping proposals as "realistic," the truth was they were inordinately pessimistic. By developing worst-case scenarios regarding world politics, systems analysis seemed predisposed to recommending only extreme, apocalyptic solutions to achieving peace.

President Dwight Eisenhower, a Republican, had warned in his January 1961 farewell address of the "unwarranted influence" of the "military-industrial complex." Meanwhile, Democrat John F. Kennedy had accessed information compiled by RAND researchers to help fuel his campaign against Richard Nixon during the 1960 presidential election. During the campaign, Kennedy regularly cited America's defense deficiencies as highlighted in 1957's Gaither Report, named after its originator, RAND chairman H. Rowan Gaither. What became known as the "missile gap," the belief that the Soviets were building nuclear missiles at a far faster rate that the Americans, was a valuable tool in the Democrats' quest to defeat Nixon.

Never mind that an investigation initiated by Eisenhower a few years earlier revealed that the Gaither Committee had based its report on faulty data. A freshly inaugurated Kennedy was sold on RAND's rational approach and its groundings in the intellectual elite.

Once Kennedy was in office, he appointed Robert McNamara his new secretary of defense. While McNamara was not a RAND alumnus, his cool, numbers-based approach to problem solving—honed in his former position as Ford Motor Company's CEO—was in perfect alignment with the RANDian worldview. Indeed, during World War II, he had performed statistical analysis for General Curtis LeMay, later one of RAND's founders, that resulted in the American firebombing of Japanese cities. Once ensconced in the Kennedy cabinet, McNamara hired Charles Hitch, head of RAND's economic division and author of *The Economics of Defense Spending in the Nuclear Age*, as his deputy.

McNamara's Whiz Kids, as the preppy defense secretary and his team became known, quickly began to reorganize the Defense Department to reflect what they saw as a new age of warfare. The air force sustained drastic cuts while the navy's Polaris submarine program received new focus and resources. Decisions were made with new thriftiness, efficiency, and flexibility. Meanwhile, the Whiz Kids developed an insouciant confidence that offended an entrenched military. Then came the disastrous Bay of Pigs invasion in April 1961, which was planned under Eisenhower but okayed by McNamara in the early days of the Kennedy administration. After CIA-trained Cubans failed to overthrow the Castro government, McNamara concluded that "the government should never start anything unless it could be finished, or the government was willing to face the consequences of failure."

What emerged was the strategy of counterinsurgency, a secretive web of tactics that included the establishment of the Green Berets and paramilitary forces tucked in areas of Latin America and Asia.

After narrowly diffusing the Cuban Missile Crisis in October 1962 through pure luck—McNamara's suggestion to defuse US missiles in Turkey before invading Cuba instead spurred a behind-the-scenes withdrawal of Soviet missiles from Cuba—the young and inexperienced defense secretary ramped up the implementation of systems analysis as the primary tool of warfare. "Every quantitative measurement we have show we are winning this war," said McNamara after visiting South Vietnam for the first time in April 1962. According to his statistics, the war would be over in three to four years.

By the time he was terminated by President Johnson in November 1967, McNamara was a broken, disillusioned man. Haunted by the mounting American deaths in Vietnam, he became convinced that the war was futile, though he would not reveal his personal thoughts until decades later. McNamara's rational, intellectual, and, dare one say, creative approach to the Vietnam War did little to account for the bloody human toll.

Decades later, some of these same pessimistic ideas about reality would come to be grouped under the umbrella term neoconservatism, which had its roots in RAND's approach to national defense. Under President George W. Bush these ideas would reach the height of their influence, as cabinet members Paul Wolfowitz and Richard Perle, two protégés of Albert Wohlstetter, the influential RAND analyst in the 1950s who later taught at the University of Chicago, played a large role in the Bush administration's decision to invade Iraq in 2003. And both hardliners helped transform US defense policy from a largely bureaucratic endeavor to a personalized one that reflected the crypto-intellectualized riffs of a few individuals who began experimenting, or tinkering, with their wildest foreign policy theories in real time.

In place of the free-range thought processes that had always made American innovation so powerful and impactful, RAND had substituted a kind of cold, amoral stubbornness. By deciding to be collectively transformative, RAND's analysts succeeded only in being collectively transgresssional. In its dyspeptic approach, RAND boiled the humanity out of one of the most innate of human instincts—vision—and muddled the true, wild-man contributions of the American individualist unbridled.

It would take decades for the idea of team-coordinated innovation to regain zeitgeist status. In its new iteration, new attention would be placed on the value of the lone tinkerer. And Nathan Myhrvold, a product of that idea factory known as Microsoft, would come up with one of the more novel approaches.

CHAPTER 5

MYHRVOLD'S MAGIC TINKERING FACTORY

T HE NOTION OF STANDARDIZING or commercializing the process of tinkering is not a new one. After all, Edison had something like that in mind from the time he built his first lab. But the exigencies of the modern world have created a whole new notion of what tinkering can do to help fuel our economy. Venture capital firms, investment operations that fund start-ups, focus on existing companies. This is something one step removed from the direct sponsorship of tinkering.

In 2000, Nathan Myhrvold sought to fill this gap with the launching of Intellectual Ventures, which he calls the world's first "invention capital" firm. Myhrvold may be best known as Microsoft's former chief technology officer and as a phenomenally rich man who has pursued a variety of interests to their logical extremes.

Intellectual Ventures is Myhrvold's effort to bridge the gap between inventive tinkering and the realization of those ideas. The firm, which Myhrvold cofounded, he said, "is about investing in inventions. So we either do our own tinkering or we invest in other people's tinkering. Either way, we are all about the process of invention."

The way Intellectual Ventures works, according to Myhrvold, sounds a little like a conventional venture capital firm but with a major twist. The company raises funds from large corporations interested in investing in innovation. Over the past decade, it has amassed a pot of more than $5 billion from a group of big names in the technology industry, including Microsoft, Intel, Apple, SAP, Nvidia, eBay, and Sony, along with some investment firms, such as Charles River Ventures. But instead of using those funds to invest in promising start-ups, it hired a lot of smart people, many of them biotechnologists, physicists, and engineers, to come up with new ideas for commercial products and patent those ideas. Then Intellectual Ventures would market and license the patents to interested corporations. With eight hundred employees and the seventh-largest portfolio of patents in the world, the company certainly wields power.

The big question is whether it represents the future of tinkering. Myhrvold views his firm as a defender of individual inventors against major corporate interests, even as some critics have suggested he is holding corporations for ransom.

Speaking with Myhrvold about Intellectual Ventures, one quickly gets the feeling it is the next step in a lifetime of tinkering that has been proven out over and over again in what can only be described as a remarkably unusual and fortunate life.

Myhrvold was born in 1959 and grew up in Santa Monica, California. Raised by his mother, a schoolteacher, he graduated from high school at fourteen and later attended UCLA, where he received a bachelor's degree in mathematics and a master's in geophysics and space physics. By twenty-three, he had earned another master's in mathematical economics, and a PhD in theoretical physics, both from Princeton.

As brilliant as he appeared, Myhrvold seemed like an unlikely hire for Microsoft. He was, by most measures, a hyperacademic type not at all interested in less lofty notions such as commerce. The title of his dissertation at Princeton was "Vistas in a Curved Space-Time Quantum Field Theory," a dead giveaway. "When I was writing my thesis at Princeton in theoretical physics, I started using one of the early computers—this was one even before the IBM PC," Myhrvold told me.

He used an early word-processing software called Magic Wand designed for a microcomputer, a predecessor to the personal computer. Then, in the early 1980s, as more powerful PCs, like the Commodore 64 and the NEC PC-98, were beginning to come out, he began writing scientific software designed to calculate complex mathematical problems as well as to create visual models of mathematical concepts. What he was attempting was similar to what later came out as Wolfram's Mathematica program.

Myhrvold and a group of friends from Princeton got involved enough in computer programming that he took a leave of absence after his first year at Cambridge University, where he had begun a postdoctoral fellowship in 1984 under the guidance of Stephen Hawking, to return to the United States. Before Myhrvold and his pals designed their mathematical software, however, the group decided they needed to build operating system extensions to facilitate the integration of their software into existing computer frameworks such as DOS.

Those operating systems extensions formed the basis of the earliest Windows system. When the friends started a company to market their operating system, Myhrvold was made chief executive of the resulting company, Dynamical Systems Research, as his leave of absence from Cambridge became longer and longer. Nonetheless, motivated by solidarity with his friends, he moved to the Berkeley, California, area in the next year and ran the company, which sought to produce a software product called Mondrian, which provided a multitasking environment for DOS. In 1986, Microsoft, then eleven years old, saw Mondrian as a missing key element in its own Windows system and bought the company and the technology.

Myhrvold was offered a job by Bill Gates, who identified him as a superior thinker he wanted on his side, and Myhrvold finally gave up his spot at Cambridge for good. For his first four years at Microsoft, Myhrvold would report to Steve Ballmer in the operating systems division. Over the next decade, Myhrvold rose within Microsoft to become the chief technology officer. All this, despite the fact that he had no formal computer or engineering training.

Despite his success, Myhrvold has always viewed himself as an outsider. That perspective, no doubt acquired as an unpopular, nerdy kid, often has freed him to pursue what interests him as opposed to what he was expected to be interested in, and take any of those interests to its logical extreme.

"I started tinkering when I was a kid," Myhrvold said. "I took lots of things apart and then put them back together again—of course, the problem was always having extra parts at the end and wondering, Where do those things go?"

His journey into the world of molecular gastronomy, otherwise known as "modernist cooking," is a good example of this. Where classical cuisine is concerned primarily with taste and presentation, molecular gastronomy delves into the chemical and physical changes that ingredients go through during the cooking process, seeking to elevate those elements to an equal level with the more traditional concerns.

An amateur cook since childhood, Myhrvold attended the École de Cuisine La Varenne in Burgundy, France, in the early 1990s, while still employed by Microsoft. There, he became entranced by *sous vide* cooking, which involves sealing food in airtight plastic packages and immersing it in a warm bath of water for extended periods at extremely precise temperatures. The technique had been around since at least the mid-1970s, but regained popularity among top chefs some two decades later because it allowed them inordinate control over the temperature at which meat or fish was cooked, meaning their dishes would never be overcooked or too rare.

Myhrvold retired from Microsoft in 2000, at age forty, and within five years he was posting his observations about *sous vide* cooking on an online bulletin board called eGullet.org. He was drawn to cooking as another opportunity for innovation. He soon discovered there was no book published in English on the topic of *sous vide*; so he decided to write one.

The result evolved into *Modernist Cuisine*, an idiosyncratic cookbook published in 2011 based on Myhrvold's experimentation at the Cooking Lab, an offshoot of Intellectual Ventures. Retailing for $625 a copy, the twenty-four-hundred-page cooking compendium includes recipes for carbonated fruit, watermelon that looks like meat, and a macaroni and cheese that includes wheat beer, sodium nitrate, and a gelatin made from red seaweed.

Deciding to approach cooking from a scientific perspective rather than a culinary one, Myhrvold invested in millions of dollars of highly technical equipment, including an autoclave—designed to sterilize medical equipment with high-pressure steam—a rotary evaporator, and a hundred-ton hydraulic press. In 2007, he established Cooking Lab and hired a team of cooks, writers, photographers, editors, and designers to carry out his off-the-wall food-science tinkering experiments.

While the resulting six volumes, which include copious photos taken by Myhrvold and unorthodox titles such as Techniques and Equipment and Plated Dish Recipes, are nearly comprehensive in their exploration of cooking's cutting edge, they offer little for the amateur chef to replicate. Rather they stand as a tinkering achievement, the extreme example of what an individual can achieve in a specific discipline without taking the most direct route to mastering it.

At his best, Myhrvold is a champion of tinkering for tinkering's sake. By ignoring the market potential of his culinary experiments and the resulting encyclopedic tome, he freed himself to follow his passion for understanding molecular gastronomy to its limit.

One of his first tinkering projects was assembling an electronic discharge machining system, known as an EDM tool, used to shape materials with sparks generated by a pair of electrodes. In those days, magazines such as *Popular Science* and *Popular Mechanics* featured plans for things you could build that were interesting—and maybe even dangerous. "You could build some pretty wild stuff," Myhrvold said. The EDM tool was one of those projects. The idea was to use extremely high-voltage electrical current to generate sparks to etch and cut metal. When Myhrvold made his, he was around nine years old.

As a kid, Myhrvold was drawn to electronics because it seemed relatively simple. "All you needed was a soldering iron and some parts from Radio Shack," he recalled.

"I'm interested in lots of things," he said. "It's not so much that I can do anything I can put my mind to, but my mind is intrigued with lots of things that leads me to try them."

Myhrvold says he was always interested by science, but he tended to gravitate toward theoretical and mathematical things that you could just think up. "When I did projects at home that was more directly tinkering, whether it was weird cooking, electronics, or I remember I wanted to build a particle accelerator—that was too big a project for me. I made some of the parts for it. I had neither the budget nor the workshop to do much more than that."

Myhrvold is quick to add that he doesn't view tinkering and invention as synonymous. "The thing that is interesting about invention is that it is a mixture of both highly intellectual pursuits—basically creativity that happens in your head—and the connection of that to the pragmatic reality," he said. "So you frequently have a situation where your great ideas are informed by your tinkering and your physical experiments, which then inspire more ideas. You end up in this feedback loop where the ability to build and tinker and play with things really encourages you to have more ideas. Conversely the ideas encourage you to have more things to go do."

Intellectual Ventures does not actually produce anything. It might make prototypes of inventions but it will never initiate a product run. Myhrvold's notion is to construct a factory of ideas.

Myhrvold contrasts his tinkering shop with venture capital firms: "Venture capitalists invest in businesses or proto-businesses. They're interested in something that can make money relatively soon, or at least deserve more funding relatively soon. They expect you to have your idea before you come to see them. We don't. We support the inventors to have the idea in the first place. That is an enormous difference."

The product of a venture capital firm is a company that either goes public or gets acquired by another company. Myhrvold says Intellectual Ventures' output is inventions, and the patents they generate.

Myhrvold left Microsoft in 2000. For a few years, he considered the venture capital route, making some angel investments in young companies. Then he stumbled upon the idea of investing directly in inventions. Myhrvold figured he could either start the ten thousandth venture capital firm or start the first "invention capital" firm. "Of course, it depends on how you look at it," he says. "If you draw the criteria broadly enough, then Thomas Edison has a good shot at this, because he had an invention lab, he raised money from investors specifically to create inventions."

Of course, Edison didn't invest in other people's inventions. He did, however, have various inventors work for him from time to time, most famously, Nikola Tesla, the father of alternating current, who claimed to have been ripped off by his former boss in the mid-1880s after he redesigned Edison's inefficient direct-current generators and didn't receive a large payment. Most of the other inventors didn't stay with Edison very long. "Because Edison's lab was about one great man," says Myhrvold. "And everybody else was there to help the great man."

Intellectual Ventures, as Myhrvold describes it, was founded with the intention of investing both in its own inventions and in the inventions of others. "I have been successful enough, I could have hired a bunch of people to help me with my own ideas," he says. "That would have been fun for me, but it wouldn't scale. Of course, we do work on some of my ideas here. But we also work on ideas from a whole variety of other people, lots of other brilliant folks with strong personalities. We figured that was the only way we could make this thing scale, and really have something that could be a sea change for how the world does stuff."

In a practical sense, this approach requires Intellectual Ventures to recruit inventors not for their existing inventions but for what they will do in the future. While the firm entertains preexisting ideas that inventors bring to it, the main goal, according to Myhrvold, is to invite inventors to work together at the firm in what he calls "invention sessions." In the sessions, participants brainstorm about new solutions to old problems or new solutions to new problems or solutions in search of a problem.

According to Myhrvold, Intellectual Ventures has filed for patents for roughly two thousand ideas that emerged from "invention sessions," among them several related to combating malaria, including some radical new diagnostics for malaria and a device that lets you keep vaccines cold when you transport them in parts of the world without refrigeration. "The most dramatic one is a device that uses a laser to shoot mosquitoes out of the sky," says Myhrvold.

Another resultant invention is a new kind of nuclear power reactor that can burn waste as fuel. "It's attracting a lot of attention within the nuclear field, so within a couple years, we can build a prototype," he says. He describes the reactor as a long-term project that will not be production ready for at least two or three years.

Intellectual Ventures has around a hundred senior inventors that work with the firm, according to Myhrvold. About seven of those are full-time employees and another thirty are professors at universities; a dozen more are consultants or run their own companies. Then there are some retired tinkerers and assorted others. "The key to working with us," says Myhrvold, "is that you have to have a day job where you haven't sold your brain. Most companies that employ folks want their brain, and you sign something that says, 'Anything you do, belongs to the company.' That would prevent them from working with us, although sometimes we've done a deal with their company."

Myhrvold favors hiring university professors who have an itch for tinkering. He says that since most are accustomed to working under the aegis of research grants, which fund plenty of researching but very little inventing or tinkering, they are eager to share their wildest ideas. "It's about studying some problem," he said. "You can win a Nobel Prize and be a fantastic person, and yet have never invented a single thing. You may have discovered something that's really important about the world."

He views Intellectual Ventures as an opportunity for such big thinkers to test their ideas in the commercial realm, an area in which he says there is shockingly little opportunity available. Myhrvold argues that the funding for invention in today's America is "pathetic." Venture capital firms are about translating an idea into a business. There are plenty of technology companies, of course, that employ engineers, "but how often do you see someone who has inventor on their business card? Essentially, never."

The simple reason, he says, is that companies are primarily interested in developing a product, not fostering the implementation of new ideas. "They'll use the term 'R&D,'" for research and development, "but it's a tiny 'r' and big 'D,' and they don't even mention 'I,' for invention, in the process." Tinkering, in this kind of corporate context, is viewed as an extraneous waste of time.

This is due in part to the fact that invention is often a subversive act. It is a disruption of the status quo. When Edison invented the phonograph, he was trying to disrupt the development of the telephone, which itself was in the process of destroying the telegraph industry.

At its essence, the message behind tinkering is that something can be done better, oftentimes by creating something new out of whatever is lying around. But Myhrvold believes that most Americans take inventing for granted, that it is just something that happens. And because inventing is considered to be risky and outside of the mission of most people and most jobs, it is rarely incorporated into a standard business structure.

Although every new idea in our technological society started as an invention, most people give invention short shrift. Venture capitalists are happy to lavish money on other parts of the technological food chain; but almost no one showers inventors with riches until they've proven themselves.

Myhrvold believes that the United States has become willfully ignorant of the unstructured kinds of environments in which the best tinkering often takes place." America still invents things," said Myhrvold. "But along the way, lots of other areas got professionalized." It's not that companies have grown to hate tinkerers; it's that everything else around the tinkering process has grown immensely. "You're not going to get me to say that the tinkering spirit is all gone, but it has been neglected."

He, like Dean Kamen, believes that the strong financial incentives for the brightest young people to become lawyers and investment bankers have gotten in the way. Of course, some tinkerers will pursue their passion because they feel they have no other choice. And for those intrepid souls, Myhrvold says Intellectual Ventures is there to offer generous sponsorship.

Myhrvold claims that his company has channeled $315 million to inventors so far. "We're hoping we can establish a new model called 'invention capital,'" he said.

Others have another term for what Intellectual Ventures has become: a patent troll. A patent troll is loosely defined as a party that purchases or otherwise acquires patents with the intention simply of enforcing them rather than actually manufacturing or using the inventions the patents were filed to produce.

Myhrvold, not surprisingly, is unrepentant about his company's defense of its expansive patent portfolio. "The process of getting a patent is pretty important to us, because without a patent, we don't really own what we have, and people can just take it from us," he said. "There would be no point in developing it." Myhrvold says that the patent filing process has become more complicated and labor intensive than in Edison's era, "so we have to have a lot of people managing that process closely."

Myhrvold is correct to place so much emphasis on the importance of patents. Indeed, the Founding Fathers considered them important enough to include in Article I, Section 8, clause 8 of the Constitution, often referred to as the Copyright and Patent Clause, as a power assigned to Congress: "To promote the Progress of Science and useful Arts, by securing for limited Times to Authors and Inventors the exclusive Right to their respective Writings and Discoveries."

Unfortunately, some in the mainstream press, as well as a few prominent bloggers, have not viewed Intellectual Ventures' exercising of this constitutional right favorably. The National Public Radio news show *All Things Considered* ran a particularly critical story in July 2011, in which it accused Intellectual Ventures of selling patents it owned to other companies "that don't make anything" which in turn were "being used to sue companies that do."

The gist of most of the criticisms is that Intellectual Ventures has been taking in millions of dollars from investors, which include some big technology companies such as Microsoft, and using a large chunk of those funds to amass one of the largest patent portfolios on the planet. The fact that it has the potential to sue a wide swath of tech companies based on alleged infringement of its patents has created a climate of fear surrounding the company in Silicon Valley.

Myhrvold, who was interviewed for the radio piece, added a note of skepticism to the proceedings, when asked by NPR's Laura Sydell whether he was a patent troll. "Well, that's a term that has been used by people to mean someone they don't like who has patents," Myhrvold responded. "I think you would find almost anyone who stands up for their patent rights has been called a patent troll."

Myhrvold went on to explain that Intellectual Ventures' intent was to protect the independent inventor who, say, had obtained a patent for a "breakthrough idea," but who didn't "have the money or legal savvy" to prevent others from stealing his or her idea. Myhrvold's firm would buy the inventor's patent and then ensure that he or she gets paid by other companies who are infringing on the patent.

NPR tried to track down a patent holder, whom Intellectual Ventures had supposedly bailed out in such a fashion, named Chris Crawford. Crawford received Patent Number 5771354 for an online backup system in 1988. The invention, according to the patent, allowed computer users to connect to an online service via phone or Internet in order to download software rentals and purchases, or to back up data. Though NPR was unable to track down Crawford, they did learn that he was embroiled in litigation over his patent, which was no longer owned by Intellectual Ventures.

As NPR later learned, Intellectual Ventures had sold Crawford's patent to a firm called Oasis Research, LLC, based in Marshall, Texas. Sydell visited the company's office, to find that it's a small office that appears to be unoccupied. NPR's report concludes, based on supporting evidence, that Oasis Research is a bona fide patent troll. To wit, in May 2011, Oasis Research filed a patent infringement suit against Oracle, accusing the tech titan of violating six different patents in the development, marketing, and service of Oracle On Demand, Oracle CRM On Demand, and other products. Oracle fired back that August with a countersuit against Oasis Research in Delaware federal court, with hopes of declaring the patents invalid.

Intent on pursuing the legitimacy of the patent, NPR had it analyzed by patent expert David Martin of the firm M-Cam. Martin showed 5,303 other patents covering similar ground as Crawford's had been issued during the time his was being prosecuted. It's not surprising that the US patent office was hesitant for many years to grant patents for computer software. It was thought that software was more like books or periodicals, original content to be copyrighted rather than an invention.

Somewhat less convincingly, Martin told NPR that 30 percent of US patents cover inventions that already exist. As an example, he mentioned Patent Number 6080436, titled Bread Refreshing Method. "So for example," said a cynical-sounding Martin in the broadcast, "toast becomes the thermal refreshening of a bread product." He made it sound as if the invention was clearly a hoax.

However, I looked up the actual patent document for the invention, filed by inventor Terrance F. Lenehan and granted on June 27, 2000. In fact, the invention is not toast, or even a toaster, at least in the conventional sense. The patent details "a method of refreshening a bread product by heating the bread product to a temperature between 2500° F and 4500° F. The bread products are maintained at this temperature range for a period of 3 to 90 seconds." It also includes a description of a heating device that would enable the process, most likely to be used by restaurants and other food-service establishments. Indeed, Lenehan was also issued Patent Number 6229117, for the oven he designed for his bread refreshening process. I am not commenting on the value of the process or its potential viability in the marketplace, only that it is hardly the same as making toast.

Indeed, the issue of the patents held by Intellectual Ventures and the companies it has sold patents to appears not as clear-cut as NPR portrayed it.

For example, in August 2011, *Forbes* magazine attacked Myhrvold and his firm in a feature titled "Trolling for Suckers." The basis for *Forbes*'s criticism of Intellectual Ventures was its methods of collecting and distributing investment capital and the poor financial returns it had registered so far. Specifically, the article took the firm to task for selling patents to technology companies that were also investors in Intellectual Ventures.

Myhrvold's defense of such accusations centers on his plan to create what he calls an "invention capital" market. "I believe that invention is set to become the next software: a high-value asset that will serve as the foundation for new business models, liquid markets, and investment strategies," he wrote in a lengthy essay published in the March 2010 issue of the *Harvard Business Review*. "The surprising success Intellectual Ventures has had over the past ten years convinces me that, like software, the business of invention would function better if it were separated from manufacturing and developed on its own by a strong capital market that funded and monetized inventions."

It would seem that the buying and selling of patents is exactly the idea, despite *Forbes*'s protestations. Indeed, the business magazine's main complaint seemed to be that Intellectual Ventures wasn't exploiting its patent portfolio enough.

Back in 2008, Malcolm Gladwell of the *New Yorker* sat in one of Intellectual Ventures' invention sessions for an article he was writing about the tendency for new patentable ideas to arrive in multiples; that is, simultaneously, by groups and individuals not otherwise connected to each other. The main topic of the session was apparently the latest developments in the realm of self-assembly, although the leader of the session, an electrical engineer with a law degree, admitted that "we may start out talking about refined plastics and end up talking about shoes, and that's OK." The range of topics discussed included minimally invasive surgery, a used CAT scanner Myhrvold had bought on an online auction site, and the "particular properties of bullets with tungsten cores."

This is not to say that some of the inventions that have come out of Intellectual Ventures don't have commercial promise. Still, Myhrvold makes it pretty clear that industrializing the tinkering process at its best is a scattershot endeavor.

But he also argues that there are many advantages to tinkering in the modern age, thanks to computer software that makes it easy to perform tasks such as CAD, or computer-aided design, in three dimensions. Almost anybody now can sit at their desk and in a matter of hours do what it once took a team of draftsmen days to achieve. Then there are computerized machine tools, which can take the computer drawing and automatically make the part one has designed.

Computer simulation allows tinkerers to test their ideas before they are even created to determine whether they will work. If you're assembling a prototype for a new kind of nuclear reactor, such an option becomes incredibly important. Intellectual Ventures owns a one-thousand-processor supercomputer loaded with proprietary software that allows it to develop such a reactor. Myhrvold says they wouldn't have been able to develop it otherwise, because they would have been unable to convince others (and maybe even themselves) that they were on the right track without building very expensive prototypes that could be exceptionally dangerous.

The rise of a technological society in general also encourages superior tinkering. To create its cutting-edge bug zapper, Intellectual Ventures bought various parts off of eBay from scrapped consumer electronic devices. They used lasers from Blu-ray players and a mirror galvanometer for steering the laser beam, from a laser printer. In a society that didn't have laser printers and blue-diode lasers lying around, it would have been much harder to come up with such a novel way to prevent insect-borne diseases such as malaria. "Because there is so much technology in so many things," said Myhrvold, "if you're a tinkerer in the twenty-first century, you can tinker at a level that Thomas Edison would be very envious of."

Still, tinkering with and inventing stuff is a pretty risky business. Intellectual Ventures tries to hedge its bets by getting involved in as many inventions (based on good ideas, of course) in as many different areas as possible to create a portfolio of inventions. By the luck of the draw, some of those inventions will be bad, but they won't all be bad.

Nonetheless, the inventions sponsored by Intellectual Ventures congregate around a few key categories: medical devices, solid-state electronics, energy solutions such as building better batteries. "Each of these areas has different dynamics driving it, and different sorts of inventors involved, but we work in all of them," Myhrvold said.

Perhaps the most unusual aspect of most of Intellectual Ventures' projects is that very few are virtual. Nearly all exist in the material world. The company is not spending the bulk of its resources coming up with e-commerce business models. Rather it is, according to Myhrvold, trying to solve big problems like malaria and trying to make carbon-free energy sources.

"We try to do big home-run problems that are so big that there isn't currently somebody working on them, or there is no one who is well positioned to do something about it," he said. "Because we're an independent invention factory, we want to swing for the fences and get some really big hits."

From a financing point of view, however, Intellectual Ventures is structured similarly to a venture capital firm or private equity outfit. Its legal structure is comparable and it raises funds from the usual suspects: pension funds, university endowments, and the like. Investors typically buy into a fund with a particular time frame and Intellectual Ventures gets to keep a percentage of the profits. If there are no profits, Intellectual Ventures is screwed.

Myhrvold says his company is doing well, though he warns that invention is a long-term process where you have to be pretty patient. Intellectual Ventures had raised around $5 billion by 2010, which will be spent over an extended period. So far, the company has spent $1 billion and returned a similar amount to its investors.

Clearly, Myhrvold views big problems as something external, something material and tactile. It's not that he rejects virtual tinkering—indeed, the majority of his firm's solutions to those problems are likely to be high-tech in nature—but rather his view of the world seems grounded in its physical manifestations.

This is not uncommon among people of his ilk. Myhrvold is an innovator, but he is also a performer. In practice, he is a collaborator, but he is also a subscriber to the "great man" approach of getting things done made popular by Thomas Edison. Frankly, his tangibly innovative projects make him look good because they improve our world in novel ways; it's no wonder he goes to great efforts to downplay the patent-gathering side of his business. How do you show that off to people? And then there's the ambiguity of its purpose.

This is all another way of saying that innovation that gets done in the material world is the kind most likely to get noticed. Tinkering that results in laser bug zappers and carbonated fruit is frankly more entertaining to talk about and demonstrate than something virtual and amorphous such as financial engineering. It draws attention to its presenter. It captures the fancy of the general public. Plus, there is something unambiguous about tinkering that results in "good things."

By contrast, virtual tinkering is difficult to demonstrate and the benefits of its results, in a number of prominent cases, are often difficult to ascertain. A good example of this is computer file sharing. To the millions who have enjoyed trading digital music, TV, and movie files over the last decade or so via so-called peer-to-peer services such as Napster and LimeWire, the tinkering that produced it solved a big problem: how to gain exposure to an ever-expanding body of culture knowledge. But file sharing during the same time became the bane of the many of the world's biggest copyright holders, mainly major media companies.

To complicate things further, virtual tinkering is more often done within the confines of a collaborative environment, far from any notion of fame or even individual recognition and where the greasy fingerprints of hard labor are easily erased. There is, however, one realm in which other rewards may compensate for the lack of these classic talismans. It is the internecine world of financial services: what used to be known in the physical world as Wall Street. And its activities have had a very physical impact.

CHAPTER 6

WHEN TINKERING VEERS OFF COURSE

THE BRITISH SCIENCE WRITER MATT RIDLEY famously and provocatively wrote in 2010 that "for culture to turn cumulative, ideas need to meet and mate," and that in our current networked culture, "ideas are having sex with each other more promiscuously than ever."

If that's the case, then the financial shenanigans that precipitated the economic downturn were a veritable orgy. In retrospect, many observers of American culture argued that perhaps ours was a nation that was getting too clever for its own good. If the best we could do as a society was to construct highly leveraged investment instruments of mass destruction, then maybe we shouldn't try our best.

One of the undercurrents that accompanied the financial crisis of 2008 was a sense that our reliance on complicated financial products that even the experts didn't fully understand reflected how far we'd gotten away from our traditional strength as an industrialized nation: manufacturing. Instead of tinkering with tools and machines with the purpose of making things, we'd become obsessed with making money at the cost of the nation's future well-being.

From the earliest days of his first term, President Obama made speeches that enhanced this narrative. "One of the changes I'd like to see is once again our best and brightest commit themselves to making things," he announced to students at Georgetown University in June 2009.

His comments took an increasingly populist tone by evoking a rosy past when manufacturing fueled the economy and condemning the widespread practice of importing foreign goods more affordable than anything American made. "America is still home to the most creative and most innovative businesses in the world," President Obama told employees at a century-old General Electric turbine plant in Schenectady, New York. "We've got the most productive workers in the world. America is home to inventors and dreamers and builders and creators. All of you represent people who each and every day are pioneering the technologies and discoveries that not only improve our lives, but they drive our economy."

No controversy there. But what followed took on a decidedly defensive tone. "Folks were selling a lot to us from all over the world. We've got to reverse that. We want an economy that's fueled by what we invent and what we build. We're going back to Thomas Edison's principles. We're going to build stuff and invent stuff." And, reinforcing the notion that the nation was once better and more productive than it is now, President Obama added, "I want plants like this all across America."

Obama's inspired words, many would argue, were exactly what unmoored American workers needed to hear. Except for the fact that such inspiring talk was overly simplistic and at worst delusional, as

Geoff Colvin of *Fortune* convincingly argued in a September 2011 column. Truth is, US manufacturing actually grew dramatically over the past decade and the value of the resultant products increased in value. The unfortunate truth is that the more sophisticated manufacturing becomes, the fewer workers it needs to increase productivity.

It's logical to conclude from this that manufacturing may not be the best place to find the jobs of the future. The most educated and privileged people in American society seemed to intuit this fact at least a decade or so earlier than most, as many abandoned the corporate world for the once sleepy world of finance. No longer would the best and brightest seek to run companies that made things; now they would commit their formidable brainpower to a world of concepts.

What rarely was mentioned was the role that tinkering played in money making. The main reason, I believe, is that this was virtual tinkering of the highest order. It also was ingenuity that ultimately caused a lot of financial pain for many innocent bystanders, people who had invested in their faith in the American dream only to find themselves drowning in what appeared to be the grandest of Ponzi schemes. In other words, this was tinkering that was destructive rather than constructive.

Because tinkering, as defined in this book, involves solving a problem with whatever is at hand, it would seem financial engineering, arguably one of the United States' most significant contributions to contemporary society, doesn't fit the bill. After all, the complex, arcane products of banks and investment vehicles fashioned by some of America's most fertile minds did not seem to emanate from a place of passion. That is, unless the desire to earn huge heaps of cash can be said to be, in any way, soulful.

It also cannot be said that this particular brand of tinkering was born without a clear purpose. The stated goal was to eradicate volatility in the financial markets. The architects of the complex web of collateralized mortgage obligations (CMOs)—sophisticated investment vehicles composed from slices of groups of home mortgages—that set the stage for the nation's financial misery were pitched to investors as an extremely sophisticated solution to an age-old problem: how to manage the risk inherent in the investment process.

But the characteristic that, in my mind, qualifies this corner of American know-how as tinkering is its role as a disruptive force. Many of the financial concepts put into play prior to the economic crisis had existed for many years, but never before were they applied with such precision. The newfound precision, enabled by computer technology and a passion for problem solving, transformed the concept of risk to such an extent that, for a brief period, very intelligent people became convinced that the age-old rules of finance simply didn't apply. And perhaps for the first time, virtual tinkering played a significant role in shifting the course of history.

Derivatives. That one word sent a chill through the average American consumer in late 2008. Needlessly complex and faintly understood, even to the analysts at the major ratings agencies who gave them their stamp of approval, credit derivatives came to represent everything that had gone wrong with our economic system and financial values.

For the average person, that is when the term "credit default swaps" came into the mainstream. Credit default swaps, also known as CDSs, didn't create anything, goes the common wisdom. Rather they and other financial derivatives destroyed massive value and cratered the American economy, without regard for those in our society who actually still make something.

But the reality is that credit default swaps were a result of an exceptionally brilliant spate of tinkering. Indeed, they were invented to solve a problem, not to create one.

It all began with the "Morgan mafia." A group of young bankers at J. P. Morgan were frustrated. They were some of the best minds on

Wall Street in the early 1990s. Derivatives, the financial world's equivalent of insurance policies, had become the ultimate intellectual challenge. Many had stepped away from other career paths to pursue the creativity finance suddenly had to offer. But unlike in other fields, finance did not offer protection for innovators and tinkerers. Unlike other engineers, financial engineers cannot patent their inventions or prevent others from stealing their ideas.

Peter Hancock, an ambitious J. P. Morgan banker from England who ran its derivatives department from the age of twenty-nine, wanted to be an inventor and took science courses at Oxford University before landing in the financial world in the late 1980s. His gravitation to derivatives seemed like a natural occurrence since they were relatively new at that time, complex in both their construction and use, and suitably obscure to entrance a young, academically inclined mind. As the head of the derivatives department, Hancock would walk the trading floor, constantly tossing out what his team labeled "Come to Planet Pluto" ideas, because they were sometimes so nutty that they seemed to fly in from outer space.

Derivatives were more of a theory than an actuality back then, allowing for plenty of creativity and experimentation. As is now well known, derivatives were novel in that they allowed investors to capitalize on the value of a variety of asset classes such as stocks, bonds, commodities, and cast, without actually having to own the assets themselves. Highly educated investors saw the value in hedging the risk of their portfolios by buying derivatives, as well as the potential benefits of taking on more risk via derivatives for a shot at outsize gains. Banks liked them because they were something new to sell and to profit from. The young financial whizzes who toyed with them, creating ever more arcane permutations, loved them because they were intellectually challenging, almost ethereal and, at least, to anyone outside of their immediate circle, inscrutable.

Derivatives are a contract between two parties regarding the price of an asset. An owner of stock shares, for example, can take out a contract to sell them around the current price if he thinks the price of the stock will fall. The advantage of owning derivatives is that the investor doesn't have to sell his shares if he thinks the price will indeed fall. Thanks to the contract, the shares will be sold automatically at the higher price if the price actually does drop. But if the price rises instead, the investor still owns the shares.

But the use of derivatives that created trouble for the economy involved buying derivatives contracts for assets that investors *didn't* own. Speculators can borrow money to buy derivatives that bet on a rise in the price of particular stock or other asset without ever owning the stock itself. If the stock price ascends, the rewards can be substantial. If it plummets (or if the speculator misunderstands the math behind the contract), buckets of money are inevitably owed.

If you're still skeptical about the level of tinkering that went into the creation of this fantastical realm of space-age financial products, it's worth knowing more about the culture that spawned them.

Hancock, though determinedly cerebral, also fancied himself a student of experimental management. First, he reorganized his team so that the sales staff could quote the price on deals without consulting the traders who put them together. Later on, he brought in a social anthropologist to assess J. P. Morgan's corporate culture. Then he polled the entire firm to figure out which departments worked most closely together and then set up a system to encourage more interaction and sharing of ideas. He also styled the makeup of a group inside the derivatives team called Investor Derivatives Marketing. Marketing certainly was one of this group's functions, but it more often served as an incubator for new structured-finance concepts.

Hancock and some of his colleagues first began brainstorming about derivatives in 1994, during a J. P. Morgan retreat in Boca Raton, Florida. They were looking to solve a few of their problems simultaneously.

The first problem was the financial industry's policies regarding innovation. Since innovations in the finance world (one of the few ways to spur rapid growth in this traditionally change-averse industry is to invent a new investment product to sell to a bank's customers) were easy for rivals to copy, the only way to prevent such rampant thievery was to come up with a new product that was essentially impossible to copy. That meant devising a new financial instrument that virtually no one could understand but that everybody wanted.

The second problem was more specific to J. P. Morgan. The legendary bank's stock price had been suffering because Wall Street didn't like its way of making commercial loans. The traditional commercial-loan business was a relationship business, and J. P. Morgan's bankers had the best relationships in the industry. Unfortunately, that meant that they were likely to grant virtually any loan they encountered, including ones they expected wouldn't be profitable. This only increased J. P. Morgan's exposure to risk. This second problem became increasingly problematic in light of the Asian financial crisis of 1997.

Hancock locked his team in a hotel conference room in Boca Raton to come up with some fresh ideas to solve both of these problems while growing the successful global derivatives business he now headed. Lots of ideas were thrown around, but the most compelling involved taking the derivatives concept and applying it to credit. The threat that a creditor might default on a loan had always been a risk. The idea behind credit derivatives—or credit default swaps, as they became known—was to bet on whether bonds or loans would default, thus taking some of the edge off a negative outcome. If the owner of the credit default swap bet correctly, he or she could profit handsomely, even if the loan defaulted.

Morgan had marketed itself to Wall Street on the notion that its commercial loans would ultimately reap high gains for the bank. But the behavior of its Asian debtors under duress erased any chance of that happening. These loans would not have the 20 percent profit margins that Morgan had said they would to its investors. Furthermore, the bank had made way too many of them, exposing it to a risk profile of exponential proportions.

J. P. Morgan needed to do something to rein in that risk. It also needed capital to divert to more profitable enterprises. But terminating loans with its longtime customers was not an option. In a business built on relationships, the bank couldn't afford to burn any bridges.

This is where Hancock and one of his bright young analysts came in. The deputy's name was William Demchak, and he was a tinkerer if ever there was one.

When it came to cooking up the formula for credit derivatives, Peter Hancock was the big thinker and Bill Demchak was the technician. Hancock asked Demchak to run Investor Derivatives Marketing. He wanted Demchak to pull together a team to make his vision a reality. The motley derivatives bunch was spread among J. P. Morgan's London and New York offices, and included a British graduate of the London School of Economics as well as a New York trader and a woman who had grown up in rural Louisiana but later studied math at the Massachusetts Institute of Technology on scholarship.

However, Demchak's key colleague on the project, known at Morgan at the time as the Credit Transformation, was Blythe Masters, a young, middle-class blond British woman armed with an economics degree from Cambridge University and a penchant for horses. Masters had started on the commodities desk in J. P. Morgan's London office, but after attending the Boca Raton gathering, sensed an opportunity. Still in her midtwenties and married with a young child, Masters nonetheless moved to New York to participate in what she viewed as the opportunity of a lifetime.

Masters later recalled that the environment at J. P. Morgan derivatives group at the time was a unique one. Instead of the alpha dog, testosterone-fueled i-banking culture of lore, Hancock and Demchak encouraged teamwork over individual achievement. While Masters had no trouble generating a constant stream of ideas for the project, she favored the collegial environment her superiors fostered, one where innovation was emphasized over personal gains, though making money was still the primary driver. It was a rare crucible for the kind of financial tinkering that happens only once a generation, if that.

Masters rapidly became something of a proselytizer for credit derivatives. She would give talks to her colleagues about their potential virtues; she enjoyed debating the finer points of credit derivatives strategy. Her passion for brainstorming rained down on the group. When Exxon faced the possibility of $5 billion in fines due to the *Valdez* oil spill in 1993, the fossil fuel giant opened a $4.8 billion line of credit with J. P. Morgan and Barclays. In fall 1994, Masters convinced the European Bank for Reconstruction and Development (EBRD) that she could dispose of the credit risk associated with the Exxon loan without actually selling off the loan, which would have offended Exxon, a longtime Morgan customer.

In August 1996, the Federal Reserve indicated it would permit banks to carry lower reserve funds if they employed credit derivatives to offset their loan risk, providing further incentive for a bank such as J. P. Morgan to push the limit in terms of ingenuity. And ingenuity was what Demchak and his team in New York were after.

Together, Demchak and Masters arrived at the big breakthrough. The key idea was to combine credit derivatives with the securitization process and create a new product that allowed the risk associated with a group of loans to be sold to another party as a bond. The name of the new product was BISTRO, for broad index secured trust offering. The point was to scrub the bank's balance sheet of the risk associated with particular loans. This was done by gathering a bunch of commercial loans together and then dicing them up into tranches, or baskets, that separated the loans from their risk. This was made possible by the fact that the loans were all pooled together and categorized by their levels of risk and return.

Demchak, an expert in structured finance, organized the loans in such a way that the resulting investment was not tied to specific loans, which made the resultant product an attractive one to investors wary about being tied down to specific real estate properties. Then the repackaged risk was sold to a shell company called a special purpose vehicle, or SPV, which in turn issued the bonds that were sold to investors. On paper and on computer screens, these new products made perfect sense, at least in concept. Without realizing it, Demchak and his team had found a solution to an age-old problem in finance: how to increase returns while minimizing risk.

Demchak and Masters spent much of 1997 convincing regulators and the ratings agencies that BISTRO was airtight. For the bankers at J. P. Morgan, plenty was at stake—approval of BISTRO was a big win for the bank. Whereas banks were ordinarily limited by international banking rules in the amounts they could lend, BISTRO freed up their capital. Normally, they needed to have a certain percentage of their loans in reserves to protect themselves from excessive defaults. But since BISTRO eliminated the bank's risk of default by selling it to outside parties, they no longer needed to hold the reserves typically required.

Launched in December 1997, BISTRO was an instant hit, selling out in just two weeks. Investors were happy to bet on the risk of loan default as long as the price of the bonds appeared cheap as compared to the amount of risk they were shouldering. The biggest customers were other banks and insurance companies, which allowed J. P. Morgan to offload $9.7 billion in credit risk, freeing up capital for other activities and drastically reducing its debt profile.

Of course, the credit risk had only disappeared purely in accounting terms. After all, it was not as if J. P. Morgan stopped extending loans to businesses at this point. If anything, it increased the debt flow since it now had fewer restrictions on how it allocated capital. Furthermore, the whole process did little to question the process of whom the bank lent money to and whether those companies were likely to default.

Indeed, had the tale of BISTRO ended here, the financial brilliance and ingenuity of Bill Demchak, Blythe Masters, and their colleagues likely would have been uncontested. After all, their jobs were to find new ways to make money for their employer and their customers. And the credit default swaps were just another innovative way of doing that.

I find it difficult to argue that the creation of BISTRO was anything but a classic tale of American tinkering. Think about it: it matches all the criteria, point for point. There was a big problem that needed solving (an excess of credit risk) and a passionate team of wizards eager to apply the tools that existed at hand to create something new. And it benefited the common good, at least at first.

BISTRO "was the most sublime piece of financial engineering that was ever developed. It was breathtaking in terms of beauty and elegance," Satyajit Das, an authority on derivatives and risk management, told Portfolio.com's Jesse Eisinger a decade later. But "in many ways," Das acknowledged, "J. P. Morgan created Frankenstein's monster."

Oddly enough, BISTRO may be one the best examples of the deeply probing tinkering of the sort that I described earlier in this book, despite the unfortunate end result it produced. Tinkering, these days, does not demand purity of purpose. Frankly, the benefits of tinkering are rarely as clear-cut as they were as recently as a century ago. In the throes of the industrial revolution, it seemed every new technological innovation was revelatory and essential to the betterment of mankind. Few could convincingly contest the value of the lightbulb or the phonograph or the automobile, for that matter, since each added a dimension to human existence so dramatic and game changing. But many of those technological itches have been scratched in the passing decades.

Furthermore, the industrial age has brought upon us the detrimental after-effects of all that great tinkering. Pollution and landfill and global warming slowly suffocate our better visions of ourselves and remind us that human ingenuity will always fall just short of solving our mortal predicament.

But as far as American tinkering goes, the CDS market fit the traditional story arc. Against the odds, a group of bright and resourceful innovators convinced others to reassess their notion of risk. This is not to say that making money wasn't a factor in this font of financial wizardry. For many involved, of course, it was the guiding factor. But that shouldn't necessarily detract from the brilliance of tinkering that filled the investment needs of two consenting parties.

The rise of collateralized debt obligations, or CDOs, however—kind of the McNuggets version of the CDS market—showed what tinkering unfettered by a societal purpose can do. In the wake of the bursting Internet bubble of 2000, the early BISTRO deals offered hope that there was still money to be made in the conceptual ether. As J. P. Morgan prospered from its newly concocted innovations, other financial institutions began to funnel their resources into the credit derivatives market. Goldman Sachs, Morgan Stanley, and Lehman Brothers were the earliest converts; even the tradition-bound Deutsche Bank built itself a credit derivatives operation, viewing it as an opportunity to break into the American markets in a way that stock and bond offerings didn't.

Part of the problem with the credit default swap gambit was that the tinkering continued well after the benefits had tapered off, at least to most people in the society at large. Otherwise, let's face it: credit derivatives, as originally devised and implemented by Bill Demchak and Blythe Masters, ultimately served to benefit both banks and consumers. Indeed, they still do.

Hedging is a tried and true investment strategy that ultimately protects nearly all investors making bets on a narrow group of investments that may collapse when that sector of the economy collapses. Both sides involved in a credit default swap want something that the other has, but neither wants to sell the credit asset underlying the trade. Many perfectly legitimate entities invest in CDSs: lenders, customers, anyone who seeks to protect themselves from the unimaginable, mainly a failure of borrowers of all stripes to pay their debts.

A 2009 survey done by the International Swaps and Derivatives Association revealed that 94 percent of the five hundred largest global companies employ derivatives, while more than 70 percent of the US-based nonbank corporations use interest rate or currency derivatives. Among US-based banking firms, all do interest rate and currency swaps, while 88 percent participate in credit default swaps.

However, the promiscuousness of the ideas that produced this situation suggests that not all tinkering is created equal. Much like in the physical world, virtual tinkering can get lost in the weeds. Ultimately, physical tinkering must produce a material thing, an object that ultimately must survive on its own merits. Virtual tinkering is not held to the same standard, by dint of the fact that a physical product is not mandatory. And yet the accountability that is a given in the material world is difficult to apply in the virtual one.

It's no surprise that in the aftermath of the CDO implosion, many white-collar workers, including those working in finance, began to question the value of their contribution to society. Around this time, when layoffs at Wall Street firms had reached an all-time peak, many former bank executives took stock of their self-worth, and the appeal of working with one's hands suddenly came back into vogue.

The timing couldn't have been better for a new wave of physical tinkerers, who had been practicing their craft away from the limelight, hoping someday to be appreciated.

CHAPTER 7

THE TINKERER ARCHETYPE IS REBORN

THERE MAY NO LONGER BE SUCH A THING as the lone, humble inventor in the United States, but the very existence of Australian-born transplant Saul Griffith at least challenges the premise that individual tinkering genius cannot flourish in our soil. Raised in Sydney and educated in material sciences at the University of South Wales with a master's degree from the University of Sydney, he arrived in America on a scholarship to the Massachusetts Institute of Technology in 2004, where he earned a PhD in programmable assembly and self-replicating machines, which sounds confusing until you learn about some of the things Griffith has spent his time doing since then. Through pondering some big issues, he has come up with an astonishing number of clever technological innovations—from a kite that tows boats to an electricity-assisted adult cargo tricycle to cheap insulation inspired by origami. The best known, perhaps, is the one that helped win him the $30,000 Lemelson-MIT Student Prize in 2004: a small desktop machine that allows an operator with little training to make a cheap pair of eyeglass lenses on demand. Griffith's motivating idea was to make glasses more affordable for people in impoverished countries. His brilliant solution worked. What he didn't realize, however, was that once an expensive lens factory is built, the cost of manufacturing and shipping a pair of eyeglasses only costs a few dollars.

Still, Griffith remained a solid idea-generating force at MIT. As a student, Griffith invented a portable electric generator that a user swings around his head to produce energy, a concept he adapted from an Aboriginal musical instrument called a bullroarer. Another device created three-dimensional chocolate objects from digital renditions. In 2007, he was awarded a MacArthur Foundation "genius grant," which was accompanied by a $500,000 prize. Griffith, in his typically thrifty style, sunk the bulk of his winnings back into his business enterprises. Many of these have sprung out of the inventors workshop, known as Squid Labs, that he established in California with a group of friends, some of whom he met at MIT. In its three years of existence, Squid Labs operated out of a warehouse in Emeryville under the slogan "We're not a think tank, we're a do tank."

A free-form ramshackle business incubator of sorts, Squid Labs produced a flurry of start-ups, including Howtoons, a website stocked with cartoons meant to teach children how to build things; Instructables, a clearinghouse of low-priced plans for a wide range of do-it-yourself projects; MonkeyLectric, a manufacturer of artistically striking lighting products for bicycles; and Makani Power, which designs airborne wind turbines meant to capture the energy from high-altitude winds unreachable by turbines mounted on towers.

Each of these creations inspired or anticipated an innovative mini-movement of its own, and together they confirmed Griffith's powers as a tinkering futurist. Spanning from practical to fantastical, they also harness a bit of the cocky, quicksilver energy that seems to be lacking in the work of many of today's innovators.

Griffith's journey to the United States was originally fueled by his interest in environmentally beneficial innovation. This is not surprising when considering that Australia is regarded as ground zero of the earth's global-warming time bomb, a cauldron of weather extremes exacerbated by an economy deeply dependent on coal as both a leading source of energy and the country's main export. At Squid Labs, he devised something they called electronically sensed rope that includes built-in sensors and conductive fibers that adjust the flexibility of the rope based on the amount of weight it is supporting. Squid also developed a power source for the One Laptop per Child nonprofit that provides children in developing countries with affordable computers. And all of Griffith's endeavors seem to be imbued with a twisted sense of humor one doesn't expect from such a prodigious innovator: one do-it-yourself project on Instructables.com includes step-by-step directions to construct a computer mouse (hardware) out of a real dead mouse carcass (referred to on the site as wetware).

Growing up in Sydney as the son of a textile engineer and university professor and his wife, an artist and weaver, Griffith's earliest memories of tinkering involve weaving machines and large manual looms that have more to do with the past than the future of innovation. Both his father and mother had at-home studios, so Griffith's childhood was filled with taking apart lots of different kinds of machines and putting them back together. "It was just a culture of 'don't be scared of any machine,'" he told me during an extended conversation we had via Skype. "I grew up around machines that weighed two or three tons, and I wasn't afraid to play with them."

An early project of his own creation was constructing a grappling hook like the ones used by Spider-Man and Batman. That single task occupied the whole of one summer. Griffith said he spent each day trying every piece of string and every piece of metal in the house "until something would stick when I threw it up on the roof of the house."

There was also a tradition in his family of making Christmas and birthday presents. Typically, they'd be either crafts projects or art projects, or else they'd be more practical items such as coat racks or camera tripods.

Griffith's fascination with his mother's weaving and knitting looms led to an interest in computers, more specifically the electronic computerization of printmaking. One of his first big mechanical endeavors was helping his father electrify one of his mother's nineteenth-century lithographic presses. "That was probably one of the first times I was exposed to real engineering, with tolerances and measurements and selecting the right motors and gears," he said.

But perhaps more critical to Griffith's development than the exposure to so many objects to tinker with, however, was the seamless connection these experiences made between the arts and science. "I think there's such an artificial division between the arts and the sciences, it's hard for me to understand," he said. "The best scientists I know are all good writers or good artists."

After working in various capacities for Australia's largest steel and aluminum producers after graduation, Griffith realized that there wasn't much else in the way of inventing going on in his native country unless it involved a better way of getting iron ore out of the ground. While living in Zimbabwe with a girlfriend, he read an article in *Wired* magazine about the decade–long quest to develop an electronic book. Fired up by the notion of all the natural resources and energy e-books would save, he became determined to be a part of the development process. "I had been working on projects in municipal solid waste treatment, and I was aware of how much we threw out in landfill was newsprint—52 percent or 54 percent in 1997. I've always had environmentalist leanings and that seemed like a terrible thing to do, and if you could eliminate newsprint with the electronic book, that would be perfect."

So he approached the professor at the MIT Media Lab who was working on electronic ink, Joseph Jacobson (also founder of the pioneering company E Ink), and told him he knew how to build a printing press but otherwise was totally unqualified in any other way to contribute to Jacobson's innovative project. "It just so happened that they were trying to figure out how to do the roll-to-roll printing of electronics" in which circuits, thin-film transistors and sometimes even semiconductors are printed on large roll of plastic or metal, so he said, 'Okay, come in.' So I sort of entered through the tradesman's entrance."

When Griffith got to MIT, the project he had arrived for, electronic ink, was pretty much finished; and its creators had left the university and formed the company E Ink to market the fruit of their labors. Within a few years, the result became readily available in electronic readers such as the Amazon Kindle. But there were still plenty of other problems to work on. The material for the e-ink display was only half the problem. To really make electronic paper come true, you have to make a world in which electronic ink will be printable on everything from flexible displays to changeable posters to clothing. The display itself doesn't cost much; it's the transistors and diodes and logic to run the display that are expensive.

When he arrived at MIT, Griffith rapidly became part of an academic culture that he describes as being unparalleled worldwide. It took a trip to Harvard for Griffith to discover what he liked so much about MIT. He recalled taking a class at Harvard Business School where latecomers were browbeaten by a professor who demanded to know why anything in the world was as important as being on time for his class. The assumption was that the late student was goofing off and not focusing on his or her studies. Back at MIT, when a student was late for electrical engineering professor Jerry Sussman's class, Griffith said, "The first question [from Sussman] was, 'Obviously you're doing something more interesting than this class: tell us all what it is, so we can all be part of it and give you better ideas and suggestions.'"

He said that his studies at MIT made him more rigorous and a bigger risk-taker in his tinkering. While he praises the American graduate educational system as being the best in the world, he says that the best academic programs understand that "the really exciting stuff happens completely externally from the curriculum and the classes and everything else."

Griffith is often asked by officials in Asian countries he visits, "How can we spur more innovation among our young people?" His answer, "Deliver free pizza to a well-equipped workshop," is not the answer most expect. But he is a staunch believer in providing would-be innovators with the resources they need to solve a particular problem and then giving them the freedom to do whatever they want.

"I still don't think of myself as an inventor," said Griffith, "although my wife loves me to write that on the passport applications as my occupation. To me, 'inventor' sounds pretentious." His sense is that if you fit the stereotype of an inventor, "then you're mentally unstable." He prefers to think of himself as an engineer well trained in science or a scientist who is "very applied" in what he does. He describes the "anarchic freedom" of MIT with a fondness that others reserve for their loved ones.

Griffith is also a strong believer in the power of teams of tinkerers, as opposed to the classic image of the "great man" inventor. "If I can implore you to do only one thing with your book, it's to kill off this one-hundred-year-old damaging stereotype," he said.

It's not that Griffith thinks individuals don't have great ideas. It's just that he puts greater value in what happens when those ideas cross-pollinate with those of others. "Because innovation happens in groups of creative people who fertilize each other and encourage each other and compete with each other," he said. "And I've never seen any innovation that's at all interesting happen without that."

Griffith's view appears to be in stark contrast with the world of tinkering established by Thomas Edison, the quintessential "great man," but in practice, it's not all that different. While Edison was the main idea generator in his lab, he had multiple assistants helping him reject the ones that didn't work and refine the ones that did. The passage of time has also made collaboration a more crucial element of the tinkering process. Edison lived at the dawn of the modern technological age; more than a hundred years later, innovating at a high level usually takes a group effort. Griffith said that all the lone inventors these days are inventing "perpetual motion machines and garden hoses," whereas the people who are doing what most consider to be real innovation are large groups of smart people with differing skill sets who appreciate each other's skill sets and complement each other.

In his description of how his work on printable electronics at MIT progressed, Griffith sketched an image of a motley crew of talented technicians with a wide range of backgrounds and abilities all laboring toward a common goal. "My background was rebuilding seventeenth-century, Rembrandt-era printing presses," he said. "And we had a chemist whose background was in doing industrial agricultural chemicals. We had a physicist whose background was in laser optics. And we had a mechanical engineer who was just extremely good at building robots." Griffith's point seems to be that it is virtually impossible to determine whose background is the most relevant to creating something that has never existed before. In the case of the MIT Media Lab, it was, in his words, "a completely unlikely cast of characters but exactly the set of skills required to make gravure-printed and nanotechnology-printed electronics" that ultimately resulted in the process needed to generate true innovation.

Citing an example outside his own personal experience, Griffith explains that the scientists who pioneered the discipline of synthetic biology—in which the building blocks of genetic matter are reconfigured to create new chemicals and drugs, and potentially, new forms of life—were a civil engineer (Jay Keasling at the University of California at Berkeley) and the physicist James Collins, who also invented vibrating insoles that help senior citizens maintain their balance. "There's not a biologist among the people who are leading the field of synthetic biology," he said.

Indeed, Griffith averred, it is often the people who seem the least qualified to tackle a specific problem that, upon combining their knowledge and skill sets, are able to devise a novel solution to a previously unsolvable problem. That is not to say that working in a team that is tinkering toward a commercial end is easy. Griffith is all too familiar with the personal conflicts that can sideline an otherwise successful project. "I'm more and more choosing the 'no-asshole' rule," he said, chuckling. "No matter how smart they are, no matter how useful they are, to make things really happen in the world, the 'no-asshole' rule is the most important rule to follow."

Griffith sees two constant threads in his own tinkering: the first is the quest for new products to solve the world's environmental problems. This category includes his solutions for municipal solid waste, electronic paper, and high-altitude wind power. The second category lies at the intersection of materials science, the study of materials and their properties, and information technology (again, the comparison to Edison holds, since the telephone and phonograph fit in this category). "Everything I've done is some union of those two things," he said.

The second of Griffith's passions strikes me as the more intriguing of the two, not to negate the incalculable value of endeavoring to save the planet from its (primarily) manmade demise. Rather it is that the second realm, in which Griffith proposes that all matter has an ability to "compute," has such broad-reaching implications from a tinkering standpoint that it manages to incorporate all environmentally conscious innovation into its sphere. A computer is nothing more than a programmable machine, something that can be made to automatically carry out a string of mathematical or logical operations. In the realm of materials science, matter is engineered based on its specific properties, oftentimes determined on a molecular and cellular level. The cutting edge of materials engineering, as being practiced at institutions such as MIT, thus involves tinkering with the cellular makeup of cells, genes, and other microscopic building blocks in order to reconfigure them to become, in essence, programmable machines.

The discipline of synthetic biology offers a good example of this process. Artemisinin, a derivative of *Artemisia annua*, otherwise known as "sweet wormwood," is the most effective cure for malaria, which is contracted by around 500 million people in third-world countries each year; and kills nearly 1 million, many of them children. As effective as it was, the demand far exceeded supply. In the early 2000s, Jay Keasling and his colleagues at the University of California at Berkeley hit upon the idea of manufacturing a cell from the genetic parts of other organisms that worked as a living microscopic machine that produced artemisinin in amounts that far exceeded those available by natural means.

After an infusion of $42.6 million in the form of a grant from the Bill and Melinda Gates Foundation, Keasling cofounded a company called Amyris Biotechnologies (now known as Amyris, Inc.), which in less than a decade was able to bump up the amount of artemisinic acid produced by each cell by a million times what a natural cell can produce. The cost of the treatment simultaneously had been reduced from nearly $10 to under $1. Mass production and distribution of the synthetically generated artemisinin began ramping up for 2012.

In Griffith's holistic approach to tinkering or innovating or inventing (in this sort of context, all three processes are ultimately involved), *any* material has the capability to be tweaked by humans to exhibit computerlike qualities. As he explains it, water and air pressure and biology all have the ability to process information, and with a bit of ingenuity, can operate in tandem with other systems to operate, at the very least, more efficiently and possibly even contribute some logic to a larger aim. Griffith likes to remind people that the first computers were rods and levers made out of the leftover parts of a jacquard weaving loom. Anything can be programmed to compute, by his estimation.

The reason this view is important is because it offers some nearly irrefutable evidence that virtual tinkering is equivalent to traditional manual tinkering, and that perhaps the most valuable tinkering going forward will be a hybrid of the two. "We have to virtualize a lot of our goods and services in order to reduce energy consumption and so, in some respects, it's great that a lot of good brains are going there," Griffith said. Viewed through the prism of Griffith's dual focus on environmental concerns and materials engineering, virtual tinkering, whether it be financial engineering or the internal logic of Google's search engine, is a way for contemporary civilization to shrink its carbon footprint while increasing the ability for humans to fashion new devices and structures with radically increased productivity from the materials that already exist.

Griffith's green values permeate virtually every aspect of life. He mentioned how pleased he was that he and I were conversing via Skype, because of all the energy saved (presumably in that it precluded me from having to hop on a plane and fly from the East Coast to the West Coast). "I wish that Skype would improve on the same improvement curve that Google is improving on," he said.

He recommends that more incentives be put in place for innovators to work in the green space, whether it be noncarbon energy (including nuclear energy) or more efficient vehicles.

Griffith's suggestion is yet more evidence of how much the world has changed in the past one hundred years. At the turn of the twentieth century, independent research labs were relatively commonplace. And in each of those workshops, "people had one of every tool that was the best of that era," said Griffith. Such workshops are no longer possible to assemble, "partly because there are so many tools to have." The impossibility of being able to assemble a comprehensive workshop means that innovation, in contemporary terms, has become a lot more difficult.

These days, the innovation centers are industrial research labs and government-sponsored research labs and university research labs, all of which are good by Griffith's estimation, "but my least favorite are the government research labs" because "we've atrophied there pretty well." Some corporate research labs produce good work, but Griffith says "there aren't enough ten-or twelve-person shops in the country."

As far as Griffith's ideal work situation, he said he most enjoys working on teams with half a dozen to two dozen people on "hard projects." But he adds, with a rare note of pessimism, that he and most of the other modern-day tinkerers he mentioned rarely spend more than 25 percent of their time doing what they are best at. A good portion of the rest of their waking hours are spent scaring up the fiscal and political resources necessary to make their ideas a reality. Not that it's ever been much different. "Leonardo certainly had to suck a lot of corporate cock to get where he was," Griffith said bluntly.

Even his notion of successful tinkering is team oriented. He doesn't necessarily believe that the commercial success of a single innovative product is as important as the influence the product has on other innovators. He uses Dean Kamen's most famous invention as an example. "In ten years' time, we'll look at the Segway as genius," Griffith said. "We're in this trough of, what the fuck is that thing? And it makes you look like a retard. But that is the right type of thing for urban transportation solutions. Unfortunately, Dean's head was a little too far out in the future. But it has heavily influenced an awful lot of things. Toyota and European companies are starting to think about vehicles that way." He also cites his own work at Makani Power in a similar framework: "It pulled a lot of people in that direction."

On the other hand, Griffith is eager to dispel the notion that the United States is the world's only culture with a strong heritage of tinkering. "I don't think it's unique," he told me bluntly over Skype, as he changed the diaper of his infant son, Huxley. "I would argue that South Africa and Australia are very similar in terms of the ethos."

Griffith thinks that America has simply built up its two-hundred-plus-year history of tinkerers into a national legend to a greater extent than these other countries. He acknowledges that the United States has had its healthy share of major innovators and inventors, but no more, he suspects, than any other frontier nation where things needed to get done and people were at least five hundred miles away from the nearest necessary tools.

In an even more acrid assessment, Griffith told me that he believed the surge in American tinkering in the post–World War II period was "really the success of military funding, it's not the success of anything else. Across every sector of the military, including NASA, this country has put more money into engineers since World War II, not only in toto, but proportionally, ten to one over the rest of the world."

Griffith also disputes the notion that the United States no longer manufactures anything. He said not all manufacturing has fled America and not the "highest-tech" and "most profoundly difficult" manufacturing.

"The US can afford to fund the craziest research," he said. There simply isn't enough venture capital to bankroll "far-out high-risk research" in Australia. Griffith, now in his late thirties, chalks it up to one of the many differences between the Australian and American cultures. Another difference is that Australia seems to trail the United

States by a few decades in terms of tinkering trends—not that that's necessarily a bad thing. For example, according to Griffith, more than 2 million Australians still tinker with their cars, even newer ones that have state-of-the-art computers inside. In the United States, however, most gear heads have evolved into custom-car nuts, rather than wrangle with the sealed plastic box that is most of to-day's car engines. Griffith said Australians don't have any fear of the processing-chip-laden cars of today.

"'Tinkerer' is an odd word," he said. "It's sort of what people who do it professionally think is an insult." He says that he thinks the biggest hurdle facing young American tinkerers in the current climate is student debt, not lack of innovative ideas or skills. "Your smartest twenty-four-year-olds have got a quarter of a million in debt just around the time when they're starting to think about wives and families or husbands and families. What choice are you going to take? You've got to take that Wall Street job. You owe the world a quarter of a million bucks for an MIT and Stanford education. Your innovation will go where you pay the best minds to go."

A defiant Griffith says he has always pursued what interests him over what makes money. "I'm interested in the energy problem," he said. "I don't understand why everyone isn't scared shitless by climate change and the energy problems coming up." He encourages young tinkerers to engage the big world problems that similarly impassion them.

Griffith has a point. Considering how quickly the business world evolves in the current era, there's no guarantee that the big money is headed where it used to. A career in the finance world, as an example, was regarded as a reliable, stable path to a big paycheck in the 1980s and 1990s. But after the meltdown of 2008, bankers were buffeted by torrents of layoffs with little hope of the storm subsiding.

And Americans routinely gripe about our inability to compete in a global economy where the bulk of manufacturing jobs have been shipped to nations like China. But while that certainly was true in the 1990s, the pendulum has begun to swing back a little, at least to a point where the argument has gotten blurred.

A sign of the shift: Google, based in Mountain View, California, decided in early 2012 to manufacture its new Nexus Q wireless home media player at a factory in nearby San Jose, just fifteen minutes from its headquarters. With manufacturing wages rising rapidly in China, it is no longer an automatic decision to assemble electronic components in Asia. Indeed, by April 2012, around one third of American companies with greater than $1 billion in revenues had plans or had considered returning their manufacturing operations to the United States, according to research done by the Boston Consulting Group.

Griffith said feelings of innovation insecurity exist virtually everywhere that he visits around the world for speaking engagements. "In Australia, people say they wish they had the British education system and the American innovation system," he said. "You're in Britain and they say, we wish we had the American education system and the German innovation system. You're in Germany, and they say, we wish we had the Australian innovation system and the Japanese education system. Every country in the world is suddenly paranoid they have lost their advantage to someone else."

Despite Griffith's fascination with the pregnant possibility of the new, he is quick to acknowledge that the paranoia nations exhibit when it comes to the future of innovation is just that: paranoia. "The number of things that are genuinely new are amazingly close to zero," he said. A healthy tinkering culture, he argues, is transparent about that reality, and uses it as way to connect the potential of tomorrow with the accomplishments of yesteryear. "We were building inverted pendulums at the turn of the last century. Henry Bessemer was building gyroscopically stabilized monorails a century ago."

Frantically tinkering to develop the new, new thing should constantly create friction with the innovators of the past. That's an advantage that American tinkerers have over all others; when we create something new and practical, we acknowledge it with an odd combination of hubris and humility that many have done something quite similar before us.

And with that acknowledgement, we fit the fruits of our labor into the long history of American exceptionalism, for better or for worse.

CHAPTER 8

PARC AND THE POWER
OF THE GROUP

TRUE TINKERING IS ALL ABOUT RISK and unusual behavior. The far-flung fanaticism that world-class tinkering requires rarely thrives in an institutional frame work. As noncorporate and free-wheeling as the world Saul Griffith describes is, it has its roots in what is perhaps the prime example of institutionalized tinkering. Of all the corporate research facilities established over the last fifty years, the PARC Corporation, an innovation laboratory born out of Xerox, is often mentioned as the one that held the most promise. And yet PARC, an acronym fo the Palo Alto Research Center, also has come to represent the best evidence that corporate tinkering in all its shapes and forms is ultimately doomed to stagnancy and failure. The very nature of corporations, and American corporations in particular, is to minimize risk and behavior that stands out. Research and development at those corporations tends to be narrow in focus and product oriented. Yet in its heyday, PARC exuded a unique frontier spirit that rarely shows its head in today's metrics-minded boardrooms. Understanding what happened and didn't happen at PARC over the past forty or so years is important in assessing the viability of tinkering in a corporate framework going forward.

Established by Xerox in 1969, PARC is best known for developing the first personal computer, which Xerox then promptly ignored, leaving the field wide open for Apple and IBM to rip off its prototypes and make billions from the results. PARC made its name by attracting the best engineers and scientists in their respective fields and allowing them to express their creativity in as wild and radical ways as they could without concern for their corporate parent.

In the wake of the Xerox 914 copier, at that point the most successful industrial product in history, the company, then based in Rochester, New York, decided to acquire Scientific Data Systems (SDS), a scientific computer concern in Southern California. Newly anointed chief executive C. Peter McColough, unfamiliar with the nascent computer industry, was looking for an acquisition to allow Xerox to compete with IBM and others in the business data processing sector.

He did not consult Xerox's engineers before offering $918 million in stock for SDS. In May 1969, Xerox shareholders approved the purchase and ushered in the modern computer era. Operating out of the old Encyclopaedia Britannica building at 3180 Porter Avenue in Palo Alto, PARC began its history with the unusual organizational move of establishing three separate divisions, despite employing only a handful of staff. PARC's first director, George E. Pake, a former physics professor and provost of Washington University in St. Louis, felt strongly that the research center should have a computer science laboratory, a systems science laboratory, which was developing the world's first laser-equipped computer printer, and a general science laboratory. By putting computer science on the same level as the traditional hard sciences, Pake and Xerox's chief scientist, Jacob "Jack" Goldman, hoped to encourage the scientists at SDS to rise to the level of innovation and productivity that had helped Xerox to become a leader in the photocopier market. The idea was to populate the relatively new science of computers with the methods and rigor of classic scientific inquiry.

Initially set up more like a university than a corporate research department, PARC was populated mostly with former academics who had never before experienced a genuine corporate culture. Thanks to a recent cutback in military spending due to the political backlash from the Vietnam War and a brutal recession, Xerox had its pick of the top research and engineering talent of the day. What distinguished PARC from other industrial development departments was its lack of clear purpose, by design. That and premium salaries that easily surpassed those offered by even the most profligate universities.

In the early 1970s, computer science did not have the pedigree it does today, so the idea of offering $30,000 to $35,000 in salaries to computer geeks with PhDs was pretty much unprecedented. But by 1970, there was not even a consensus among the nation's top computer scientists that devising a computer with a high-powered display for personal use was a worthwhile goal. The prevailing model at the time involved building the largest computer technically feasible and allowing only for professional computer operators to time-share on the unit.

A handful of fortuitous developments made PARC a catalyst for change in this environment. The first had its roots in the US Defense Department's Advanced Research Projects Agency, or ARPA (and renamed DARPA in 1972), originally formed to create new missile technology in the national panic that followed the launch of *Sputnik* in 1957 by the Soviet Union. By the early 1960s, the space program had been pulled out from under the military umbrella and given its own agency, NASA, leaving ARPA to concentrate on civilian scientific research. Despite the fact that ARPA's mission was supposed to be related to national defense, with ample funding to suggest it was something of a governmental priority, it lacked a clear mandate.

J. C. R. Licklider, the former behavioral psychologist who first headed ARPA, suggested that the world's largest user of computers, the Defense Department, should fund a world-class computer science research program. The result was the Information Processing Techniques Office, with a then astronomical $14 million budget (worth roughly $100 million in today's dollars) and none of the bureaucratic red tape that other federal agencies had to deal with. While Licklider funded mostly large-scale time-sharing computer projects during the mid-1960s, his successor, Bob Taylor, was keen on directing ARPA's capital into smaller projects, the most prominent being Project Genie, conducted at the University of California at Berkeley. Project Genie's goal was to construct a computer system for use by ten to twenty operators simultaneously, rather than the hundreds required for larger-scale systems. The theory was that a smaller, more affordable computer could be distributed more widely and thus empower more users overall.

The core of the Genie system was the SDS 930, made by Scientific Data Systems, which retailed for $73,000. It was later upgraded with about $5,000 of additional hardware and sold as the SDS 940, for $173,000 per unit. It became one of SDS's best-selling products.

But again, SDS's strength was scientific computing, not business computing, which had somehow eluded Xerox's Connecticut-based management team. When PARC's engineers recommended purchasing a PDP-10, made by the Digital Equipment Corporation and rapidly becoming the computer of choice for research laboratories nationwide, the purchase order was declined by management on the East Coast, based on the stubborn but incorrect belief that the SDS 940 could be modified to match the abilities of the PDP-10.

In a desperate move, the computer geeks in Palo Alto decided to make their own version of the PDP-10. Armed with PARC's unique creative philosophy, crafted by Bob Taylor—that everything they designed should be designed for everyday use—a core group of scientists and engineers swiftly entered a heretofore unimaginable environment in which they were granted the authority to erect their own computer system to their own specifications.

The second key factor in PARC's ascent was the hiring of Alan Kay, a rumpled, eccentric computer scientist, erstwhile jazz guitarist, and acolyte of Seymour Papert, inventor of the LOGO educational programming language. LOGO was designed to teach children about computers by allowing them to see the immediate effect of typing simple programming commands that would appear on a screen and move a toy robot around the floor.

Kay, often cited as the archetypal computer nerd, arrived at PARC in 1970. Kay was a new kind of computer scientist who did not fit the stereotype of the previous era, a timid, clean-cut Poindexter in a lab coat. Kay, to the contrary, had a wild shock of curly hair and a moustache, accompanied by a swagger that somehow injected the computer world with a shot of coolness that it never really recovered from. Prior to being hired at PARC, while still a graduate student at the University of Utah, he had envisioned a device he called the Dynabook, even going so far as to build a nonworking model of it; it looked remarkably like a cross between an Amazon Kindle and a laptop computer. Kay was attracted to the job at PARC because some of the computer whizzes he admired had recently taken positions in the rapidly expanding Palo Alto lab.

The third development, in 1971, was the emergence of silicon semiconductors, introduced by Intel, which quickly replaced the bulky ferrite core memory that had been the industry standard since the early 1950s.

The combined force of these factors resulted in the MAXC, Xerox's answer to the PDP-10. The actuality of the MAXC, which at the time had the largest semiconductor memory of any computer on earth, was ultimately less important than the team and methods that created it. The MAXC took eighteen months to deploy, in contrast to the decades it had taken for most computers of the time to be assembled.

For all the potential success that the MAXC represented, it became a source of conflict between its designers at PARC and the Xerox executives back east. Xerox's corporate officers typically viewed technological change as something to be monitored in order to protect the company's business plan. They were interested in predicting future trends in a general sense, in order to understand what the world was going to be like. That way, Xerox could defend its existing product line against impending competitors. On the other hand, Alan Kay and his colleagues saw only innovation and opportunity ahead. Kay famously stated that "the best way to predict the future is to invent it."

It is that faith in serendipity that made PARC's accomplishments so impressive over the next decade. When headquarters demanded a concrete plan for what was ahead, PARC delivered a folder with seven reports, each solely written by a PARC scientist, outlining what he hoped to accomplish.

The vision outlined in that folder showed remarkable prescience. From portable flat-screens to CD-ROM-like photo optical media, the reports described innovations that are now humdrummedly mainstream but at the time seemed nothing short of futuristic. Over the next decade, as it failed to capitalize on nearly all of the innovations it clearly anticipated, PARC came to embody all that was wrong with the corporatization of American tinkering. Certainly, it is clear from chronicles of the era, that was not the fault of Alan Kay and his merry band of programmers.

In a fascinating article by Stewart Brand, the founder of the *Whole Earth Catalog*, that ran in *Rolling Stone* in December 1972, Kay, Taylor, and others at PARC were portrayed as staggeringly brash, knowledge-fueled hippies determined to wrest control of computers from the hands of stern corporate factotums who were only interested in the technology's value as a high-tech abacus. Their esprit de corps was based less on some political notion or ideological cant than on the idea that computers could only achieve their full potential in the hands of individuals, not corporations.

But PARC's style of group tinkering was perhaps its most valuable asset and contribution to the annals of innovation. One of the best examples of the form was known as "Beat the Dealer," or just "Dealer," after the book called *Beat the Dealer* by Edward O. Thorp, a professor at MIT who devised a card-counting system to win at blackjack. PARC's version involved twenty or so of its researchers assembled on mustard-colored beanbag chairs as one of their ranks pitched a new project and the rest tried to find its flaws. The presenter, or dealer, though left to his own defenses to support his idea, had the advantage of being allowed to set the rules of the debate as well as the topic. One dealer used his time to show how to disassemble a bicycle and apply the proper lubricant. Another discussed at length how similar computer algorithms were to cooking recipes.

Along a similar line of reasoning grew a collaborative process known as "Tom Sawyering," after the enterprising protagonist from the Twain novel. The concept here was that when a researcher came up with an idea for a new project or device, he would try to make it a reality by rallying those who were interested to help put it together. If the project began to show promise, the informal team would work on it for the next six months or so; but if it failed to gel, the participants could slowly migrate back to their own work and the project would simply dissolve.

In late 1972, PARC's most infamous product would emerge from this ragtag tinkering process and forever change the course of computing for good. Alan Kay had spent the earlier part of the year sketching out a simple programming language for his modified Dynabook project, now known as the miniCOM, and instructed his engineers to make it a reality, which he named Smalltalk.

Kay's follow-up proposal, to build a personal computer that ran Smalltalk, was met with chilly disdain by the suits at Xerox headquarters. So Kay waited until the executive responsible for the computer lab's budget was out of the office for a couple of months on a special task force and then told his team to build the personal computer as fast as they could.

The computer's design began in November 1972 and was completed by February 1973. All of those dealer meetings and Tom Sawyering sessions finally paid off, allowing the engineers to pull bits of knowledge from problems they had already explored and deploy them in building a new prototype. In classic tinkering fashion, the engineers assembled the new device from stray parts they had lying around. They repurposed memory boards that had been built for MAXC; the display monitors were from another project, a large networked system called POLOS that had more to do with the old world of massive shared computers than the compact individual units they were now creating.

The result was the Alto, the world's first personal computer. The Alto included a screen about the same dimensions as a letter-size piece of paper held vertically. The monitor was large and bulky by today's standards, but otherwise the Alto was way ahead of its time—it had a mouse that moved a cursor on a screen that showed exactly what the user would be printing. And while hardly the most powerful or fastest computer around, the Alto had other virtues, the most significant being the freedom it allowed operators, who could now sit at desks wherever they liked rather than in isolated rooms filled with cumbersome mainframes. Xerox manufactured around two thousand Altos in the coming months.

While the Alto was never introduced as a commercial product, it earned Xerox a place in the annals of computer history, though not exactly the one its creators at PARC had envisioned. The reason was the result of one soon to be famous visitor to the Xerox offices in West Hollywood, California.

His name was Steve Jobs.

By the summer of 1979, Jobs's company, Apple Computer, was already a presence on the West Coast computer industry landscape. The Apple II had already been released and the fledgling company was readying itself for an initial public offering. At this point, Xerox and the folks at PARC had little awareness of Jobs and his distinctly countercultural operation. As for Jobs, he was unlikely to consider a partnership with Xerox, due to his suspicions of large, faceless corporations.

But then an Apple engineer told him about the project some of his buddies at PARC had been working on, and Jobs was intrigued. So when Xerox requested participation in the last round of financing at Apple before the public offering, he made them an offer. He agreed to sell the company 100,000 private shares of Apple at $10.50 per share in exchange for a simple visit to PARC's research lab and an explanatory tour. Confident in their company's supremacy, Xerox's executives believed they had made a ridiculously advantageous deal. As developers of one of the largest and most powerful computers on the market, they regarded Apple as a manufacturer of technology for hobbyists. Besides, the PARC researchers had shown the Alto and Smalltalk to representatives from the Central Intelligence Agency with little consequence.

Little did Xerox suspect what Jobs's true intentions were. Apple had been developing Lisa, its follow up to the Apple II, but Jobs had been dissatisfied with some of the more user-friendly aspects of it. After a number of skirmishes over how much of what they were working on they were required to show him, the PARC engineers demonstrated their graphical user interface, a new kind of computer interface that used graphic images instead of words, including a series of overlapping "windows," which could be dynamically moved with a small, rounded pointing device known as a "mouse." They also revealed that they could scroll text on the screen as if it were a piece of paper.

Jobs was blown away. In 1980, he requested a license to use Smalltalk in the Lisa. Xerox refused to grant it, having already cashed out its investment in Apple. So Jobs hired away one of Smalltalk's creators, Larry Tesler, who would become a key developer of the Lisa and Macintosh computers, eventually rising to the position of Apple's chief technology officer.

Life at PARC was never the same. Despite a recent decision to expand its research budget significantly, Xerox did not capitalize on the personal computer interface it developed. Instead, it tunneled millions into the development of silicon-based integrated circuits, despite the fact that such circuits were already readily available elsewhere.

Of course, PARC invented other technologies that would prove influential in the not so distant future, including the laser printer, Ethernet networking, the optical disc, and LCD technologies, among others. Unfortunately, Xerox was never properly able to capitalize on most of those innovations, either. The laser printer market was first developed by IBM in the mid-1970s, after Xerox delayed the sale of its own 9700 printer for five years while it endlessly debated the cost-analysis merits of doing so. Cisco Systems and 3Com later cornered the network hardware business, due to Xerox's initial instinct to keep the technology secret, followed by its decision to license it to anybody for a one-time license fee of $1,000.

But the real consequence of the squandering of PARC's tinkering capital was the effect it had on corporate research in the decades after. No large corporation today would permit its research arm to develop technology that was not tied to a specific, anticipated product. Not even Xerox. It seems counterintuitive that PARC's success in producing new technologies, even if Xerox couldn't properly capitalize on them, didn't convince other large companies of the value of building their own internal innovation labs. But the risk was perceived as too great, and in most cases, relegated to the world of much smaller start-ups.

Since 2002, PARC has existed as a wholly owned subsidiary company of Xerox rather than simply a research arm. And while Xerox accounts for around 50 percent of its business, it has other clients including Samsung, NEC, and VMware. Many of the engineers and scientists who worked at PARC in the early days went on to populate then unknown companies such as Apple and Microsoft. PARC's impact continues to be felt, but the lessons drawn from its failures continue to weigh on the corporate version of American tinkering.

Henry Chesbrough of UC Berkeley's Center for Open Innovation has argued that PARC's biggest problem was that it nurtured what he calls a "closed innovation paradigm" rather than an open one. PARC in the 1970s sought to create all stages of a new product's development within the confines of its own corporate structure. It wished to own not only the tinkering environment but also the facility that developed the resulting innovations into products, the factories that manufactured those products, and the teams that ultimately marketed, distributed, and serviced the products. That had been the vertical integration model used by all of the top industrial companies in the post–World War II era.

But what happened at PARC, according to Chesbrough, offers proof that another approach was in order. As he meticulously documented, nearly all of the innovations devised at PARC in its most fertile period were later produced and sold at other, smaller companies populated by former PARC employees. Among the better known of those technologies were the Macintosh computer and the Bravo word processor (which later became Microsoft Word). Of course, many of the other companies built on innovations developed at Xerox failed. But that's exactly the point. One company, no matter how large or well capitalized, is rarely equipped to handle all of the ups and downs contained in the normal course of development for technologies that are genuinely radical and new.

Xerox was not completely unmindful of the value of technologies developed at PARC; rather it decided not to develop technologies it felt were not worth its further investment. In most cases, the developers of these innovations left PARC with Xerox's blessing and only after Xerox arranged for a hefty licensing fee.

But the licensing fees could never come close to matching the revenues realized elsewhere by the products that became huge commercial successes. They also did little to capture the value of the ongoing evolution of an innovative idea. Most of the technologies that left PARC during those years were not appealing ones, at least at the time Xerox gave its okay.

Chesbrough used SynOptics as an example. Founded by two former Xerox executives, Andy Ludwick and Robert Schmidt, in the mid-1980s, the company was created to develop and market a technology that provided high-speed Ethernet service over optical cables, an innovation developed by an individual PARC researcher, that allowed for much faster transference of data. The problem was that optical cables were not yet prevalent, and building the necessary infrastructure looked to be decades away. Furthermore, users of the new fiber-optics technology would have to rewire their offices with optical cables as well, to connect computers with printers and other devices. As a consequence, Xerox decided the research was no longer financially prudent to pursue, since the mainstream consumer market was nowhere near adopting it. As a smaller, independent company, Ludwick and Schmidt were prepared to wait until the market caught up with the technology; they amicably brokered an exit from Xerox, with Xerox retaining a 15 percent interest in the new company.

Shortly after breaking free of Xerox, SynOptics discovered that its technology worked nearly as well over the existing copper-wire infrastructure, providing substantially faster transfer speeds without the need for an overhaul of the infrastructure. It was the freedom to keep tinkering without the pressures of a corporate parent that allowed SynOptics to stumble upon this new application. This radically changed the way its product was developed and dramatically accelerated the company's growth curve. Three years after its inception, SynOptics went public in October 1988. It soon was worth north of $1 billion and merged with a company that later became part of Nortel.

The original technologies developed by an inventor are often not the ones that ultimately make it to market and become commercial successes. Xerox's research and development process was designed to capitalize on new products that it could market to its existing customer base, and naturally favored ones that related to its copiers and printers. However, those were not the technologies that necessarily had the most commercial potential. Rather they were the ones that best fit into Xerox's preexisting system for developing new products.

At its heart, Xerox's PARC incubation process failed to acknowledge some of tinkering's self-evident truths. First, that innovation most often begins with the vision of an individual and it is only through support for that individual's efforts that the new technology ever makes it past the early stages of development. Second, managing the technological risks of innovation has little to do with managing the associated economic risks. And third, perhaps most important, no matter how large and well funded a corporate incubator is constructed, it is at best a crapshoot as to what the benefits of that support system will be.

In the United States, this issue is perhaps more protracted than in other nations, since the individual's urge to innovate is girded by our tinkering history. And yet Americans have built most of the biggest and strongest corporations the world has ever known. The contradiction inherent in these two realities has at times spurred the country to create great products (Apple's first iPod, which Sony failed to imagine, is a good example).

But more than we'd like to admit, it also has resulted in stagnation, especially when corporations use their firepower to bombard any new ideas they find threatening to their existing businesses. This might explain why some truly radical technological innovations of recent years have emanated from foreign shores.

CHAPTER 9

A TRIO OF ALTERNATIVE TINKERING APPROACHES

C LEARLY, TINKERING IS A KEY TRAIT of the American technological firmament, but plenty of tinkering occurs elsewhere. In the
following three parables, I explore some more recent developments I
believe offer distinct insights into the ongoing evolution of the tinkering process. Two are about European tinkerers and the other is
about a Chicago architect. Together, they illustrate a few elements I
believe are largely absent from today's tinkering landscape in the
United States.

Deep in the bowels of the gargantuan Javits Center in Manhattan, a
rumpled, professorial-looking man in a gray suit and tie spoke from

a podium in clipped, German-accented English to a room of fifty or so geeky conventiongoers on October 21, 2011. The occasion was the 131st biannual Audio Engineering Society (AES) Convention, but this keynote speech hardly seemed to garner much interest.

The speaker, who had a salt-and-pepper beard and moustache and wire-rim glasses, was Karlheinz Brandenburg, better known as the father of MP3. MP3, for those few who are unfamiliar with it, stands for MPEG-1 Layer 3. MP3 the most common digital format for compressing music so that it can easily and quickly be transported over the Internet. Without MP3 and its related formats, there would be no digital music, no Napster, no iPod, and the music industry would still be successfully selling compact discs under its old business model. By making music incredibly portable, the MP3 format pointed to a future in which information could flit around in the ether via cloud-based or wireless communications systems.

Little more than a decade ago, such concepts seemed fantastical, unlikely to have much of an impact in the near future. Now MP3s are so common and ever present that they seem almost old-fashioned, despite their relatively short history. This might explain the low attendance at Dr. Brandenburg's lecture at the AES conference. To the scientists, audio engineers, and acousticians assembled in New York, MP3 had become the ordinary, the common language, and a revolution that had come and conquered.

Although some American engineers, including James D. (JJ) Johnston of AT&T Bell Labs, played a role in the development of the MP3 format, the story of how Karlheinz Brandenburg arrived at its ultimate codec is one of painstaking tinkering that might not have been able to occur in the current-day United States. This is not because Germany was some hotbed of human knowledge but rather because the right set of tools were made available to the right set of people for an extended period of time, thanks to receptive financial backers. There was an element of luck involved, as well. But the opportunity to make something out of luck is something that can be planned as much as anything else.

The story of MP3 is very much like that—luck and skill and opportunity operating in unison. As such, it is the tale of a tinkerer's paradise.

Karlheinz Brandenburg was born in 1954 in Erlangen, Germany, the oldest of three children and the only son. From an early age, he was a classic tinkerer: he loved music, and so got an amateur radio operator license, which allowed him to communicate with others over the airwaves, and soon began to build his own amplifiers. By the middle of high school, he had launched his own business, selling amplifiers to his fellow students.

The second current of his early years was his membership in the Boy Scouts; Brandenburg was a Boy Scout for all of his childhood, and came to love participating in group activities, especially those that involved building something.

His love of music extended to playing musical instruments; he learned to play the recorder, the violin, and piano through formal lessons. He later taught himself to play guitar "for around the campfire," Brandenburg told me. After showing no exceptional talent on any of the instruments he encountered, however, he solidified his role as a devoted and attentive listener.

By the time this budding tinkerer-entrepreneur-musician arrived at the Friedrich-Alexander University of Erlangen-Nuremberg, he had become something of a pragmatist, and, thus, enrolled in the school's engineering program. A year later, however, realizing he still had a passion for numbers, he also enrolled in the university's mathematics program. He graduated with a degree in electrical engineering in 1980 and one in mathematics in 1982. But Brandenburg quickly discovered that pure theoretical mathematics required a focus and dedication he was unwilling to grant them, so when it came time to pursue studies toward a PhD, he took a definitive turn toward electrical engineering.

In 1982, while still deciding what studies he might pursue, Brandenburg was asked by his PhD advisor, Professor Dieter Seitzer, to take over work on a project he himself had been pursuing for a while: how to transfer high-fidelity music over a phone line. Seitzer's prime interest was not music but improving the sound fidelity of telephones, which use only a quarter of the sound spectrum heard by the human ear. This goal was of little use to Brandenburg; he did, however, become entranced with the idea of how much of an audio signal could be removed without the ear detecting any distortion. Seitzer was impressed when Brandenburg succeeded in completing the task the professor assigned to him.

Seitzer also had an important affiliation: when, in an effort to put Germany at the forefront of cutting-edge microelectronic systems, the Fraunhofer Society—a countrywide German research organization funded partially through the state, but primarily through government and corporate contracts—decided to establish the Fraunhofer Institute for Integrated Circuits (IIS) in Erlangen in 1985, it named Professor Seitzer as its founding director. As a result, Brandenburg almost immediately had access to a wealth of technical resources, including state-of-the-art signal-processing equipment, as well as a mission to help develop products that might one day earn the institute profits via its patents.

Over the next four years or so, he spent his time experimenting with different ways to compress sound by eliminating certain elements of the sound spectrum. The goal was to learn which parts could be removed with little or no discernable degradation of the listening experience. The higher the bit rate, or number of kilobits per second, a file uses, the higher the audio quality. Generally, the problems with music file compression arise in finding a suitable balance between file size (lower bit rates result in smaller, more portable files) and fidelity (which improves with higher bit rates and larger, more cumbersome files). Brandenburg's innovation was to find the perfect formula for smaller file size and improved audio quality.

"The standard way of doing it was with a bit-allocation algorithm," said Brandenburg, referring to the mathematical formula that helped translate a large quantity of audio inputs into a reduced set of signals, essentially eliminating certain frequencies in order to shrink the size of the file. "But what I did over time was create a much more flexible system." It was more flexible because it took into account previous knowledge of speech and musical sounds, providing a more holistic template for compressing those sounds with losing their realistic qualities.

Brandenburg used the "analysis by synthesis" approach of speech perception, which allowed him to focus on how the human ear actually hears music rather than relying on a very generic bit allocation formula. He famously used a recording of folk singer Suzanne Vega singing her hit song "Tom's Diner" to calibrate precisely the right frequencies that could be removed from a music file with noticeable loss in fidelity. "I picked 'Tom's Diner,'" he said, "because it is begins as an a cappella and then adds very metronomic rhythmic elements." (According to Brandenburg, it is very difficult to conceal deteriorated audio quality of an unadorned human voice.) He also incorporated the clever use of filter banks, which isolate and highlight key elements of an audio signal, and other high-tech frippery to concoct his compressed music.

Despite his success in completing the task he had first been given by Dieter Seitzer, an early attempt to obtain a patent for his sound compression work was met with a good deal of skepticism. The patent examiner apparently told Brandenburg "there is no high-fidelity music at 128 kilobits per second," he recalled, the common transfer rate of an MP3 digital file, meaning the distortion at that level of compression would render any recording unlistenable.

Brandenburg admitted that, as proud as he was at the time of the elegant programming that went into MP3, he was despondent about its future as a marketable music format.

With good reason. In 1983, the recording industry introduced the compact disc, a digitized-content format that could be played over and over with no noticeable degradation of audio quality. The CD, which packed ten times the fidelity of an MP3 file—1,400 kilobits per second versus 128 kilobits per second—onto a plastic disc with a diameter of 120 millimeters, seemed to have eliminated the need for an easily transportable music format such as MP3.

Thrilled to have found a way to boost the revenues of an ailing industry, the major labels priced CDs substantially higher than the vinyl records they were meant to replace. Understandably, they had little interest in a competing digital format with lower audio quality.

Indeed, it took until 1989, a full six years after Brandenburg did his pioneering MP3 work, before there was even an inkling that there might be a practical use, and a commercial market, for the industrious student's work. By then, Brandenburg was already wrapping up final work on his PhD.

That same year, Germany financed the Digital Audio Broadcast project as part of a European effort, known as Eureka 147, to develop a digital standard for broadcasting over the radio airwaves, which turned into a bake-off between the IRT (Broadcast Technology Institute) in Munich and the Fraunhofer IIS in Erlangen. Out of that challenge, which took place from 1989 to 1991, the MPEG-1 Layer 3, or MP3 format, the technology Brandenburg had developed, emerged as the best compromise between file size and sound quality, although the MP1 and MP2 formats continued to have their uses.

Also around that time, Brandenburg began a yearlong stint at AT&T Bell Labs in New Jersey, a descendant of Alexander Graham Bell's research efforts and ground zero for the United States' efforts to advance sound compression technology. He says he visited Bell Labs as an exchange postdoctoral researcher in part to see how such research operations conducted business in a country that rarely funded technological innovation to the extent his home country did. Brandenburg spent his time in New Jersey fine-tuning the MP3 codec alongside James D. Johnston, a prominent American audio engineer. He was deeply impressed by what he witnessed during the time he spent at Bell Labs. "It was like a university with famous professors," he said, "but no students." Even more surprisingly, despite its corporate underpinnings, he found the research being done at Bell Labs was "even more detached from the realities of the marketplace than at Fraunhofer."

Brandenburg was awed by the sheer number of internationally known experts that had been gathered at the facility; if he had a question about any aspect of the audio research he was working on, he could usually walk down the hall and speak with the inventor of that very piece of technology. And the methods of getting research done were strikingly similar to what he was used to at home.

But there were also some substantive differences, which hinted at why Germany had taken the lead on the MP3 technology. The first, Brandenburg noted, was that Fraunhofer's organizational structure is such that the same dozen engineers, himself included, worked on the MP3 compression project over a period of more than ten years. Fraunhofer offered engineers an extremely stable institutional framework and a collegial atmosphere, as well as access to all the latest research gear they needed to conduct their experiments. None allowed himself to be lured away by lucrative employment elsewhere, despite the fact that Fraunhofer pays salaries that are closer to academic ones than to the corporate packages that top-flight electrical engineers typically enjoy. "This was a group of people who really wanted to work together," said Brandenburg.

There was another key factor that distinguished Fraunhofer from its American counterparts. It's not that Germans are inherently less motivated by profits than Americans; rather, the German legal infrastructure favors the rights of individual innovators over those of corporations. In Germany, as in the United States, patents developed under the auspices of a corporate institution are owned by the company, not the individuals who did the work. But unlike in this country, Germany requires that employers share a portion of the royalties obtained from patents with the employees who are named on the patents. All told, Karlheinz Brandenburg's name is attached to twenty-seven United States patents, along with coinventors, which means he has been amply compensated for his contributions to the MP3 technology. He said his royalty payments—generated mostly by licensing to MP3 encoders, the virtual players that allow users to listen to MP3 files—have far exceeded his Fraunhofer salary. "Let's just say I was able to build myself a nice house without taking out a mortgage," he said. He also used some of the royalties to found a venture capital firm called Brandenburg Ventures.

Lastly, the Fraunhofer Institute's hybrid financing structure offers a unique, nationwide opportunity for science and innovation to flourish. There are sixty Fraunhofer Institutes scattered throughout Germany, focusing on a panoply of disciplines including everything from molecular biology to computer graphics. All of the institutes are run on the same model: 30 percent of the budgets derive from state and federal land grants; the other 70 percent come from governmental and industrial contracts.

The combined effect is that Fraunhofer's research facilities offer a level of institutional stability that few corporate cultures could provide while still injecting that rigorous experiment-based environment with a commerce-driven impetus for innovation.

Brandenburg noted that there were a number of gaps between the time when he first established the basic parameters of MP3 and when it was finally adapted as a common format for music compression in 1994. In a purely corporate environment, such research would likely have been abandoned over that period. But patience is a built-in virtue at most Fraunhofer outposts, since many are affiliated with nearby universities.

In 1993, Brandenburg was named head of the audio and multimedia department at the Fraunhofer Institute for Integrated Circuits in Erlangen. Then in 2000, he became a professor at the Institute for Media Technology at Ilmenau Technical University in Ilmenau, a small town closer to Berlin. When the Fraunhofer Institute for Digital Media Technology was established in 2004 in Ilmenau, he was named its first director.

The swirl of myth and awe that surrounded Thomas Edison during his lifetime no doubt played a large role in creating the image of the "great man" as innovator and genius. As the great defender of innate individualism, the United States has long propagated the notion that brilliant ideas that change the world magically pop out of brains predestined for greatness. And yet more and more examples provide evidence to the contrary. It is compelling to appreciate how many of today's great tinkerers are in fact teams of tinkerers, all laboring toward a common if sometimes fuzzy goal.

The world of video games offers some of the best examples of this kind of tinkering as team sport. *Angry Birds*, easily the biggest video game success story of the late 2000s, was created in an incubator-style tinkering environment that emerged from the University of Helsinki in Finland. In 2003, Niklas Hed, a twenty-nine-year-old Finn, entered a competition sponsored by the Finnish cell-phone corporation Nokia and Hewlett-Packard, with two friends, to create a multiplayer mobile game for one of the first smartphones made by Nokia. Their winning entry, called *King of the Cabbage World*, was one of the earliest multiplayer, real-time games that could be played together remotely by individual phone users. Impressed by the sophistication and inventiveness of the group's entry, and in particular by Hed's programming abilities, Peter Vesterbacka, then an employee of Hewlett-Packard and one of the competition's judges, recommended that Hed pursue a career in programming games. He also suggested the trio start their own video-gaming company.

And so in 2004 Hed approached his cousin Mikael, a business school graduate four years his senior, and asked if he would be the chief executive officer of his new company, then called Relude. After a year had passed, with little success raising funding, Niklas asked his uncle, Mikael's father, to contribute funds to their start-up. Kaj Hed, a successful Internet entrepreneur who had recently sold a company he had founded, agreed to kick in one million euros. Mikael soon left the company after a disagreement with his father over its business plan, leaving Niklas to run the company, now known as Rovio, on his own.

But instead of creating games for his own company, Niklas and his team of developers spent the first few years developing them for other companies such as Real Networks, Namco, and EA to make ends meet. While the company prospered and was able to hire more employees, the business model was heavily dependent on having hit games, and Rovio didn't have any. By late 2006, the future was not looking good for Rovio. Teetering near bankruptcy, Niklas laid off most of his fifty employees, whittling his staff down to a small but impactful twelve.

Then, on January 9, 2007, everything changed. As Steve Jobs emerged on a stage in San Francisco during the Macworld conference, unveiling the first iPhone and the app store that would accompany it, Niklas realized he had a golden opportunity to revive his ailing company. He lured back his cousin Mikael, and the pair set out to create a game for Apple's remarkable new device. They would continue to do development work for other gaming companies, but all their remaining hours would be devoted to their iPhone game.

Their proposed budget for the project was twenty-five thousand euros, though it would later balloon to four times that amount.

The design team established a punch list of requirements for the new game. Since the market for the iPhone was nearly everybody, it needed to have broad appeal, so it wouldn't have a science fiction or horror theme like many of their earlier titles. The game also had to work on multiple platforms, though the iPhone version would be the main focus. They also determined that it should be "physics-based," meaning it would mimic the physics of the real world, a quality that adds to the random fun factor of video games. The game would not need any instructions to get started, and a user would need to enjoy playing it as much for a minute as for an hour, which required a quick loading time. It also needed a recognizable logo to get noticed in Apple's App Store.

Rovio's head designer, Jaakko Iisalo, spent the next two years sketching out drawings of literally hundreds of characters, as the company also kept busy developing other games. Iisalo would generate ideas in groups of ten, and present screenshots of what he envisioned. It wasn't until March 2009 that he sketched an angry-looking bird that grabbed his colleagues' attention. From that point on, the team of developers worked together to create a game they all enjoyed.

It was by no means a straightforward path. Early versions included one where each bird corresponded to a matching block; when the user touched the block, the bird would take to flight and demolish it. The birds themselves did not have unique powers. The entertaining feature of being able to fling the birds across the screen came later, as did the slingshots used for the hurling. Together, they also came up with the targets of the birds' anger: pigs who thought nothing of devouring their eggs. The various colors of the birds, their otherworldly squawks, and their ability to bomb objects below with their eggs, which could be accelerated by touching the screen, were all choices the team made by gut feeling, with no market research or focus-group testing to guide them.

During this time, the tinkering process became quite fluid and organic. Since the programming team was working on at least four other games while developing *Angry Birds*, the time they spent on their pet project was discretionary and thus driven by interest. Sometimes a member of the team would be testing a feature and get caught up in playing the game for fifteen minutes or more, suddenly surrounded by a group of fellow programmers. By allowing themselves to drift naturally toward features that were the most entertaining and addictive, they knew by the time it was done that they had created something special.

When Rovio released the game in December 2009, it didn't get much attention, and its inventors feared they had met with failure once again. For the first three months after it went on sale in the App Store, *Angry Birds* gained little traction. The English-speaking markets, which the game needed to crack for access to big sales numbers, showed little interest. So instead, Rovio concentrated on much smaller markets, such as Finland, which only required a few hundred purchases to make the game number one. Next came Sweden and Denmark, followed by the Czech Republic and Greece. Once they had racked up a total of 30,000 to 40,000 downloads, they arranged to have the game distributed by Chillingo, a game publisher with a strong connection to Apple. Based on the strength of the Chillingo distribution deal, Apple decided to highlight the game on the front page of its App Store on February 11, 2010. Just for the occasion, Rovio offered a free, simplified version of the game (it normally sold for $0.99), as well as an animated YouTube clip that featured the game's characters, only the second clip produced for an iPhone game. Last time I looked, the trailer had more than 74 million views. Within three days, *Angry Birds* was Apple's number one, most downloaded app. It was later reconfigured in HD for an iPad version, which sold for $4.99.

The *Angry Birds* mania snowballed from there, hailing a new era in software sales. No longer would software primarily be purchased for a high price in a box at a store. Analysts later estimated that Rovio's revenues averaged between $50 million to $70 million. The game has been downloaded more than 50 million times, and has spawned a franchising venture as well as plans for a big-budget movie. Sales of stuffed animals based on the game's characters also skyrocketed. In March 2011, Rovio was able to raise an additional $42 million in funding from investors such as the founders of Skype and Accel Partners.

I report this information not to suggest that taking a collaborative tinkering approach to creating video games is the equivalent of the process it took to create some of the life-changing devices I have described in the previous pages. But the development cycle has certainly been challenged. The Rovio style of tinkering may be preferable to the now omnipresent product development cycle in the United States, which demands that ideas must be scientifically probed by market researchers and test panels before ever making it past the tinkering phase.

As their discipline has become increasingly concerned with function over form, and as a result boiling some of the passion out of the process, some American architects appear to have naturally evolved toward the more individualized Rovio style of tinkering. It hasn't hurt that it has become more acceptable in recent years for small boutique firms to get hired for large projects that decades ago would have automatically gone to one of the large firms such as Skidmore, Owings & Merrill. This has in some part to do with computer imaging software that allows a tiny architectural operation to produce drawings and models with the same speed and accuracy that once would have required dozens of draftsmen weeks and months to produce.

But one also suspects that clients are finding that smaller firms are more likely to take a team tinkering approach that prides creative and striking solutions to big problems that traditional firms with large cost overheads have a tougher time tackling.

One such firm is Studio Gang Architects, founded by Jeanne Gang, the 2011 recipient of a MacArthur "genius grant." The Chicago-based Studio Gang made its now well-known name on the Aqua building, an eighty-two-story residential condominium built in the up-and-coming Lakeshore East neighborhood.

The resulting tower, which was completed in 2010 and is now one of the tallest in Chicago, has been praised by critics for its undulating structure featuring concrete balconies that ripple out in irregular shapes that do not conform to the lines of the actual building. The aesthetic effect is an echo of Lake Michigan, which is only a few blocks away, but the design is also practical: the balconies act as passive solar shades for residents, who also have remarkable views of the city at heights previously unavailable at any price point in the Windy City. The building is also exceptionally environmentally friendly, with heat-resistant reflective glass installed where the balconies don't provide shade, and a rainwater collection and storage apparatus to supply sprinklers on the green roof.

Gang grew up in Belvidere, Illinois, a small town just past Chicago's suburbs. Her father was a civil engineer, and she spent her childhood trailing her dad on trips that toured rural Illinois's roads and bridges. Gang attended the University of Illinois at Urbana-Champaign and the Harvard Graduate School of Design, from which she graduated in 1993, after which she went to work for Rem Koolhaas's firm in Rotterdam.

She moved briefly to another firm in Chicago, before launching her small practice in 1998, after winning a commission to build a theater at Rock Valley College in Rockford, Illinois, not far from where she was raised. Gang clinched that first project by learning ahead of time that the dean of the college was a hydraulic engineer and that the contractor for the theater had constructed bridges. Emboldened by her knowledge, Gang proposed a theater with a folding kinetic roof and immediately hired a structural engineer to show it was feasible. She got the job soon after that, and the theater was finished in 2003.

The Studio Gang office, which includes around thirty-five employees, has also gotten a lot interest for its prevailing work style, which could only be described as team tinkering. When Stephen Zacks of *Metropolis* magazine visited in 2008, he described the design team as "surprisingly relaxed," despite the fact that at least four projects were at the schematic stage and a proposal for a high-rise residential development in Hyderabad, India, was due to leave the premises in less than an hour.

What the reporter found were teams of designers "grouped in hives of activity throughout the office," which was managed by Gang's partner and husband, Mark Schendel. Gang sat "slightly secluded" in an office off to the side, coming in only occasionally to weigh in as a collaborator.

At their essence, Gang's organically designed buildings seem to evolve out of an organically designed workplace. "I like to bounce things off people," she told *Metropolis*. "I'm less likely to sit in here and do a sketch and then deliver it. I would more likely think of an idea and go out there immediately and ask, 'What do you think of this?' I have to hear a response."

In the course of writing this book, I've had the pleasure of visiting more than a few small companies that have reconfigured their workplaces to adapt to this new perspective on tinkering and its relationship to innovation. Yes, they are mostly large, open, loftlike spaces where employees sit at desks or tables shoved together instead of cubicles or enclosed offices. But these work environments are not fashioned in the style of the dot-com companies of yore.

There are no Ping-Pong tables or massage chairs or beds installed over desks. No free food or showers where employees who never leave the premises try to revive themselves after pulling all-nighters trying to reach ever-impending deadlines.

The point of these restructured work environments isn't to bleed employees dry. Rather, it is to reinvent the notion of the creative workplace, where tinkerers and innovators are engaged with their fellow workers in developing new products in a way that better mirrors the unfettered human thought process—and allows for inevitable hiccups that accompany it.

In such environments, it is not uncommon for the employees to reconfigure where they sit based on the specifics of the project they are working on. These workplaces, not surprisingly, also tend to have flat hierarchies. Don't be surprised if the CEO sits among his or her employees; this isn't an affectation but rather a way for everyone to keep tabs on what new ideas are developing as they develop.

These newfangled workplaces lack most of what we generally associate with traditional corporate America. Workers are counseled to treat each other respectfully, and yelling is prohibited. Not every idea is a good one, but each is giving consideration by the team and accepted or rejected through a measured, collaborative process. The casual observer may notice that there's a lot of pleasurable chatter, but these aren't your average wage monkeys goofing around: these are fully engaged producers.

These hubs of innovation know what many large American corporations have yet to learn: that invention is idiosyncratic, difficult, joyful, frustrating. And that Americans, when properly nurtured and incentivized, are uniquely suited to pursue it. There are few nations in this world that simultaneously embrace both the childlike senses of awe and wonder and the Calvinist work ethic, and also have the nearly unlimited financial capital to sustain the tension such a condition creates. I can think of only one: the United States.

It didn't surprise me, therefore, though it was something of a revelation, when I stumbled upon what struck me as one of the most insightful voices regarding the nature of tinkering running a summer program in Northern California. His name is Gever Tulley.

CHAPTER 10

A DIFFERENT KIND OF SCHOOL

S EVEN OR EIGHT YEARS AGO, Gever Tulley, who is now in his late forties, finally confronted his own childhood. It happened around the time when all of his friends started having kids. Since he had no kids of his own, Tulley hadn't thought much about his own upbringing lately. He recalls a time in his early thirties visiting a friend with young kids. They were talking quietly about nothing in particular when the friend suddenly yelled at her son, who was innocently brandishing a stick, "Is that a stick? You know the rule. No playing with sticks!" He would, from that time, question the various "rules" we have for our kids, and the unintended cost of our obsession with safety.

As Tulley makes it clear, his own childhood in Mendocino, California, was different. His parents were remarkably free of the rules that bound other kids, as were those of many of his friends. "In those days, they'd boot you out the backdoor in the morning," he says. "And they'd be like, 'Be home before dark.'"

Maybe he'd go to a friend's house, maybe they'd head out on their bicycles. Nobody knew where they were or where they were going. Maybe they were out in the woods with some tools and wood, building a castle in the trees. Maybe he'd jump off a high rock into some water. "You developed a true sense of risk and danger really through a series of minor bumps and scrapes—and maybe a broken arm here and there," he told me.

Tulley does not seem scarred by those experiences. Instead, they seem to have imbued him with an eternal sense of wonderment about the world. It would be tempting to call Tulley's perspective childlike, if it weren't so complex and mature about the way it manifests itself in his current day-to-day existence. His vision may be best described as "Tinkering 2.0."

I met with Tulley in New York City. He was passing through on the way to Europe, where he was booked to reprise one of his now wildly popular TED Conference presentations. Tulley gave his first TED talk, titled "Five Dangerous Things You Should Let Your Kids Do," in March 2007. On the original list were play with fire, own a pocket knife, throw a spear, deconstruct appliances, and either break the Digital Media Copyright Act (by converting a digital music file from a paid format to the MP3 format) or drive a car. Tulley's point in each case, delivered with plenty of humorous asides, was that these taboo activities all provide valuable hands-on learning experiences for children while also building confidence and self-reliance skills. The video stream of the talk on the TED website has been viewed more than 1.5 million times.

TED talks—"TED" stands for technology, entertainment, design—are typically about innovative ideas that run counter to conventional wisdom, and are often delivered by well-known leaders in their respective fields. Past TED speakers include Bill Gates, Bill Clinton, Al

Gore, Google cofounders Sergey Brin and Larry Page, Jane Goodall, Gordon Brown, Richard Dawkins, and various Nobel Prize winners. When Tulley first delivered his talk, he was neither a leader nor well known. He was a contract computer programmer who recently had founded his first summer program for children, known as the Tinkering School.

Tulley was born in Mendocino, California, three hours north of San Francisco, to parents he describes as beatniks rather than hippies. They predated the hippies, he says, and listened mostly to jazz, not rock and roll. Mendocino was an old logging town that nearly vanished in the 1940s but underwent a revival as an artist colony around 1957, when the Mendocino Arts Center was established. Cheap land drew hippies from San Francisco up to California's north coast where communes thrived in the 1960s and 1970s, and marijuana became a major cash crop. Tulley describes his father as a "fisherman-slash-beatnik poet," someone who naturally took to the area's alternative lifestyle.

His strongest recollection of that period was how his parents' friends, a varied mix of artists and other creative types, treated him with the same respect as an adult, even when he was a young child. It left a lasting impression: there was no need to talk down to kids.

Soon after Richard Nixon was reelected president in 1972, his parents elected to move the family to the small Canadian town of Nakusp, British Columbia, as an act of protest. Eight years old at the time, Tully had a hard time making the adjustment from Mendocino's temperate climate to the harsh western Canadian winters. When foot after foot of snow fell during the winter, the roads would be closed, and school would be too. But when the lush wilderness thawed, Tulley and his brother reaped the benefits. The family's property had a creek on it where the boys would jump on an inner tube and float five miles downstream. Along that route lurked limitless adventures that brought them close to nature; they frequently saw bears, mountain lions, and other wildlife.

But it wasn't all fun and games; the inhospitable climate meant having to come up with inventive solutions to unusual problems on the fly; in the winter, the family had to build an ice dam to ensure it would have enough water for the season. The Tulley family would stay in Canada for both of Nixon's terms, during which time Tulley experienced what he considered the far superior Canadian grammar schools. By the time they returned to California in 1975, just in time for him to attend middle school, he was far ahead of his classmates. He would face several years of boredom, but with this boredom came an opportunity.

As relief from the mind-numbing review of lessons he was already familiar with, Tulley spent a lot of time "being in my own head in the classroom." He kept to himself mostly and entertained himself with his own thoughts. Tulley had always had an active inner life, and this was simply a full-time chance to indulge it. Thankfully for him, in his first year of high school, his school introduced a program known as the Community School, which offered more creative learning opportunities for unorthodox students like the one he had become, as well as what he calls "the feral children of Mendocino County," kids who needed hands-on attention from the school's educators.

It was around this time that Tulley was diagnosed with spondylolisthesis, a helical twisting of the spine due to a congenital defect. As a result, at age thirteen he began suffering painful sciatica, including a constant feeling that a hot poker was being jabbed into his left leg. He would require surgery involving light traction while four vertebrae of his spine were welded together. Afterward, Tulley was placed in a body cast that went from his knees to the top of his head; a kind of a hard-shell space suit. He was then suspended on spokes inside two giant hoops to keep him in the proper position for healing.

And so he wouldn't get bed sores, the hoops were placed in a tray with rollers that allowed him to turn himself over. The whole contraption was then put on wheels, so he could push himself around with his gloved hands.

He would be a prisoner in this horizontal rotating cage for about three months. Fortunately, a woman who worked at the Community School offered to pick him up every day and take him to school, since his parents were both working. The woman arrived each morning with a van, and with the help of a ramp, got Tulley into it and strapped his tray down in the back. Once at school, he could wander the school in this bizarre, horizontal position. "My legs were slightly spread, I had these Velcroed sweatpants that would fit around the spokes so that I was discreet in public," he says. "And that was my life at that period of time."

It was perhaps a tribute to the free-thinking outlook of the school, and students attending it, that Tulley was able to pass through a good part of the school year in this fashion. Indeed, since many of the kids in the program were permitted to take on any educational project they could conceive of, his unusual state of being was hardly remarked upon. Apparently some students mistook his ailment for some sort of wild experiment. Around this time, something clicked in Tulley's head: if he could live through this, he could get away with anything. Freed from the fear of being stared at for appearing different, he became determined to pursue his passions no matter where they led him.

It was during this otherworldly experience that Tulley believes the seeds were sown for what would become the pinnacle of his life's work, an alternative education program called the Tinkering School.

His bedrock principle was that kids could do real work. The work didn't have to be abstracted for students to understand it. Never mind reading a physics textbook; why not build a physical thing that expressed the same ideas? The same approach, Tulley decided over time, could be applied to music and art, even history and philosophy.

Around the age of six, Tulley recalls, he began asking adults visiting his family's home if he could tag along to wherever they were going. He was interested in finding out more about the things adults did. At first he got a lot of friendly stares and polite nos, but around the age of nine, people began to say yes. Among his early experiences of this kind was hitchhiking to San Francisco with an adult family friend, spending a couple of days with the friend in the city, and then hitchhiking back.

After graduating high school with a GED, due to his unorthodox schooling, Tulley applied to and was accepted to the University of California, San Diego; coming from a poor family may have helped him. Although college was his first experience with conventional educational methods in quite a while, including attending classes and taking tests, there was some question already whether it would be of any value to him. Tulley had already been making a living as a computer programmer for three years before matriculating. At fifteen, he had been hired to write code for medical devices.

Not surprisingly, college was not a good fit for Tulley. He began getting that old bored feeling again. While he enjoyed his film and creative writing classes, he experienced little else of value to him. He would last just one quarter.

He remained in San Diego while his girlfriend at the time finished out her year of college there. When she transferred to Santa Cruz, he decided to follow her. Meanwhile, Tulley quickly found a job repairing some of the first portable computers, the Kaypro and the Osborne, at a little cutting-edge computer shop that also sold typewriters. A typical problem he saw: two floppy disks shoved into the disk drive because users didn't know to remove the first one when prompted to INSERT DISK 2 by the computer. It was 1981.

For many young people, such a lowly job might have seemed like drudgery, but for Tulley it was an instant education. "When the Kaypro II came out, they'd solved a lot of the problems in that second generation," he says. "That was kind of a marvelous thing, to have been so intimate with the construction of the first Kaypro, and then to have the Kaypro II come out, and open it up and these simple changes fixing the electromechanical problems of the first generation."

Tulley later realized that while his unorthodox upbringing in Mendocino had exposed him to alternative ways of looking at the world, it didn't channel into his notion of ambition and accomplishment. "There wasn't a culture of getting out and making something of yourself," he says. "It was okay to lead a low-key, moderately productive life. Not that people were lazy, but the grand ambition was kind of uncommon. It was okay not to have it." A lot of people had moved up to the Mendocino area to escape the city and lead a semiagrarian life.

Tulley associates the drive to tinker that he and other Americans feel with the westward expansion in the United States in the mid-1800s. The exhortation by the US government in that era to get out west and finish populating this giant country had a lasting effect on the young nation. Sure, there were other motivators, such as the prospect of gold in California. But the notion of pushing farther into the unknown frontier bleeds into almost everything Americans do, whether there's a need for it or not. There's a natural tendency to believe that things could be just a little bit better. Good enough is never good enough. Americans never leave well enough alone.

Of course, not every American goes on to be a tinkerer, but the potential sits lurking in the heart of many a citizen, raised to believe that fame, prosperity, or maybe just recognition awaits those who try just a little harder.

Take the French fry, for example. Invented elsewhere (just exactly where is forever up for debate), the thin strips of deep-fried potatoes were perfected in the United States by the J. R. Simplot Company in the late 1940s. Founded by J. R. "Jack" Simplot at the age of fourteen, the Idaho-based company mass-produced a frozen fry concocted by Simplot's scientists. In 1967, Simplot agreed, in a handshake deal with Ray Kroc, to supply McDonald's with all of its frozen French fries. But the perfect French fry was not enough. More innovation was demanded. The result? Ripple-cut fries, waffle fries, curly fries, Tater Tots—the variations are endless.

"The taxonomy of French-fry-making machines is just ridiculous," say Tulley, with a chuckle.

Tinkering School began as a six-night sleepover summer camp program in 2005. Tulley's first campers were his niece and a few of his friends' kids. One child flew all the way from France to participate. Another two jetted in from Connecticut. Quite a remarkable turnout considering Tulley had no camp accreditation and the only publicity for the camp was posted on his blog. To make things more challenging, Tulley still held a full-time job as a software engineer during the first session. Then there was the infamous document parents had to sign before their children could attend. They actually had to print the words "I understand that my child may be injured or killed at this camp."

That first year of Tinkering School, the kids arrived to find a huge pile of plywood plates and two-by-four blocks. Their charge for the first day: build chairs. Tulley began by taking all the chairs out of his studio. He asked the children to take a seat. After a nervous chuckle from his campers, Tulley suggested they build some chairs. The next day, they built a twenty-foot-long truss-beam bridge that connected the studio's deck to a tree and carried the weight of the entire class. On the third day of camp, they built some towers of various heights. The tallest ones allowed the campers to climb up onto the roof of Tulley's studio.

On the last two days of the program, the kids built a working rollercoaster out of wood with the help of Tulley and few other adults. In the name of continuity, Tulley wishes he had inserted one more interim project relating specifically to wheels, but he feels the same effect was achieved indirectly. "When you look at what a roller-coaster is made of, it's basically a chair sitting on wheels riding over a series of bridges and towers," he says. "So the kids had this vocabulary of skills and what was strong, and why it's important that when you screw two blocks of wood together, that there be no gap between, because you end up with a loose joint."

Tinkering School's first curriculum supported a key tenet of tinkering: the value of building on previous inventions. Tulley contends that, due to the week's previous projects (the chair, the bridge, and the towers), the kids were better equipped to solve the problems related to erecting a rollercoaster. "It basically was two days of building and then tinkering with it to tune all the corners and things like that," he says. In the end, the campers built 120 feet of track.

The Tinkering School's days are broken up into segments, and for each successive segment, Tulley has built in some additional options, based on the assumption that different groups of children will progress through tasks in different ways. For the initial group, they could have built ladders before building towers, but Tulley decided they were sufficiently handy to skip that step. Instead of a rollercoaster, they could have built a drawbridge, in the style of London Bridge, with two towers, suspension architecture, and a moveable roadbed. "But I could tell by the kids' velocity," says Tulley, "that we could pull off the rollercoaster by Saturday morning."

One of Tulley's main guiding points for Tinkering School is that all of the projects have to be real. No fake tools or preordained conclusions. In other words, if the campers are going to build their own boats (which they did one summer), they are going to try them out in the water. If they don't float, the kids sink.

Tulley's theory is that if children get a sense that their experiences and their classroom education are following a script——in other words, that the authorities in their lives already know the conclusions——they pick up on that and it lessens their interest in the topic. That is not to imply that he has no ulterior motives behind some of the projects.

"For the bridge, I wanted to get the kids to understand why when they look at so many things in the world, they are made out of these triangles," Tulley says, sketching a few on a blank page of a notebook that he carries everywhere he goes. All the kids had were twenty-four-inch and sixteen-inch chunks of wood. After that, he moved to plywood: four inches on one end and either twenty-four or sixteen inches on the other end. "And we had four-by-four-inch two-by-four chunks. Those constraints gave them a problem space to work within." The idea is for the kids to learn how to make their project work with what they have—the essence of tinkering.

Tulley, however, swears that he had no idea how he and his campers would build the rollercoaster until they actually did.

Tulley held the Tinkering School for the first three years at his house, as the roster of campers quickly grew. Then he moved it to a farm that he rented some fifteen miles away. Seventy-five percent of the ten campers in each session paid tuition, which now is $1,450 per week, and helped fund the other 25 percent, who didn't.

Somewhere along the line, friends of Tulley recommended that he attend the annual TED Conference. The first time he went, he paid his own way, as an attendee. Then TED leader Chris Anderson sent out an invite informing guests that for half a day before the conference began, any attendee was welcomed to propose their own topic to talk to the other TED attendees about.

Tulley took the bait, and sent back a proposal. Almost off the top of his head, he suggested a talk on Five Dangerous Things You Should Let Your Children Do, for which he planned to provide numerous visual aids. He shortly got a response from Anderson, who informed Tulley that his proposal had been accepted. He still, however, had to pay the stiff conference fee of thousands of dollars (official speakers get a discount).

Tulley actually first spoke at the TED preconference on March 6, 2007, a day or two before big names like venture capitalist John Doerr, President Bill Clinton, and singer Paul Simon would deliver rousing talks to packed auditoriums. The idea was that he'd just go over a few quick entertaining ideas he had formulated while operating the Tinkering School summer program. He did the first talk in a small and only half-full room. It took a little over nine minutes. But in the two hours between the talks, word apparently got out, because he found himself moved to the largest room, and entered to a packed crowd. And that might have been the end of it, except for the coincidence that Tulley's talk was chosen to be the first TED talk videotaped and posted online, and it soon went viral. He became an Internet sensation.

Tully tried to turn his idea into a book, but he found that traditional publishers were reluctant to take a chance on an untried writer proposing a book that would, among other things, suggest that people ought to let their kids play with knives, among other worrisome activities. Frustrated, Tulley decided to self-publish with the help of his wife, Julie, and the result, *50 Dangerous Things You Should Let Your Child Do,* appeared after three months of eighteen-hour days spent writing. It was a true tinkerer's effort: Julie laid it out and designed it with the help of InDesign software. They made a deal with a print-on-demand publisher, an affiliate of Amazon.com. The price of the book was set at $25.95, because that was the lowest price they could sell it for in England without owing money on each copy. On the bright side, they earned $6 for every copy sold in the United States. The book sold 12,000 copies through Amazon, and by spring 2011, Penguin agreed to publish an expanded version of *50 Dangerous Things.*

But Tulley's goal in life is not to be a best-selling author or a highly compensated speaker. He believes the United States is squandering a national resource. He identifies it as "the creativity and innovative ability of generations of children." Tulley believes that the average American education of today has a tendency to turn out more consumers than producers of ideas. It's no coincidence, he posits, that the best memories most adults have of high school are either social events or sporting events.

A year or two after kids attend Tinkering School, Tulley often calls them up to assess their level of retention. He is continually amazed at how much former campers can recall after so much time has passed. "The detail with which they remember riding the roller-coaster or flying the hang glider that they built, or the sailboat, or the motorcycle that we made, or whatever it is, the minutiae they remember and the principles that are burned into their brains from those experiences, those are lasting, durable memories," he says.

Tulley believes that school science classes should be competing with drama classes for students' durable memory space. English class should be so full of adrenaline that we're sifting football memories from English class memories. He thinks schools should measure children's engagement with the material along with all the test scores and attendance records.

It's worth mentioning here that Tulley has no formal training as an educator. "The things that I don't know about pedagogy are in all the books that I haven't read," he quips. Rather, Tulley views his Tinkering School efforts as high-concept problem solving. Tulley has no children of his own, which perhaps gives him some distance from the issues that parents wrestle with and worry about. If anything, he sees his own interests more in line with those of kids, not intellectually, but in terms of what has the potential to engage him. He claims he could spend all day with a pile of sticks and twine, figuring out what he could make from them.

Tulley seems to be arguing that tinkering at its essence is innate, a preternatural drive to make new things from stuff that already exists. But something happens to children from the time they are in grade school until they become adults. Unfortunately, this something is sometimes called public education. Due to the trend toward quantifiable educational progress, many lower-quality school curriculums, and some higher-quality ones as well, are tilted toward how students perform on standardized tests.

Not surprisingly, Tulley expanded the Tinkering School's mission by cofounding an actual private day school in San Francisco, called Brightworks, with another alternative educator, Bryan Welch. For the 2011–2012 school year, Brightworks enrolled twenty children in first grade through seventh grade at a tuition of $19,800 for the year (though at least a third received financial assistance). The initial student body was composed with gender balance and cultural diversity in mind; the program also is designed so that younger and older children work alongside each other at times. Tulley's goal is to keep adding students each year and extending the curriculum through the twelfth grade, eventually achieving a total student population of eighty.

He shows me his notebook, which is filled with diagrams, pictograms, and a variety of arrows, wavy lines, and descriptions. In spots, these drawings look more like the plans for some kind of Rube Goldberg contraption instead of a new kind of learning experience.

Those drawings later became what is now known as the Brightworks Arc, a carefully rendered illustration on the school's website that shows the three phases of learning each student passes through in the school's program: exploration, expression, and exposition.

Tulley says one of the first inspirations for starting the school was behavior he witnessed at the Exploratorium, a science-oriented museum for kids in San Francisco. "If you just pick out an exhibit—it pretty much doesn't matter which exhibit it is—and you just watch it for twenty minutes or an hour, you'll see this recurring behavior where a child comes over and starts playing with it, and the parent comes up and starts reading the plaque on the wall and tells the kid,

'Oh, push that button, here's what we just saw, okay, let's go.'" After forty-five seconds or so, the parent is pulling the kid to the next exhibit, regardless of the child's engagement with what he or she is doing.

Tulley likens such an experience to the traditional school experience, where knowledge is meted out in forty-five-minute periods, and schools spend increasing amounts of money in an attempt to optimize those forty-five minutes to achieve the highest test results possible. He believes, however, that what is needed is a greater diversity of educational experiences rather than an excess of fine-tuning. He suggests that some kind of hybrid education, involving both traditional book learning and alternative, hands-on experiences may be the ultimate solution, since only around 7 percent of students are being served by the existing model.

In the 1970s, there was an explosion of alternative schools based on the free school and Sudbury school models, which eliminated grades, curricula, and traditional teacher-student hierarchies in the name of better learning. Tulley views those experiments as necessary steps in the evolution of pedagogy, but his school will take only certain elements from each. He admits his understanding of the historical framework of education is lacking.

For that, he relies on Bryan Welch, his Brightworks cofounder. Welch, who has a dual bachelor's degree in education and journalism from Berkeley, runs a summer program called A Curious Summer, which takes an exploratory approach to discovery and learning with an emphasis on self-discovery. Tulley posits that Welch successfully eliminates what Tulley describes as the "dictatorial role" he takes at the Tinkering School in determining what project the children will work on. For summer 2011, A Curious Summer offered a freeform weeklong exploration of sound that encouraged kids to build their own musical instruments, learn music theory, and put together their own microphones and speakers.

In the hybrid model the two men came up with, they curate a number of projects around a topic such as wind, power generation, nautical history, meteorology, or the arts. Then they take the students through an exploratory journey of the topic, during which each student naturally gravitates toward some aspect of it that interests him or her (Tulley and Welch did test runs over the previous years). Finally, the students pair up in groups of two to four to develop their own individual portfolios. As Tulley describes it, the school begins as something akin to a museum (the exploratory phase) and then transforms halfway through each project into a workshop (the tinkering phase). In this context, algebra becomes a skill needed to measure how much PVC pipe a student needs to complete a particular project. "The idea is to always contextualize those topics," says Tulley.

If the topic of the day is wind, then the children start with activities such as flying kites, experimenting with wind tunnels, and building wind turbines (exploration). Then they move on to a project of their own making, which might be anything from building a sailboat or creating a work of art featuring tornados (expression). In the final phase, the students put together a detailed presentation and deliver it to an audience of their peers and teachers. They also create a portfolio of documents and objects that creates a record of what they have learned.

Despite the nonhierarchical approach to education, the school does not eschew adult supervision. The school has a six-to-one teacher-to-student ratio, and each student has his or her own mentor, who might be a collaborator on any given project. The school staff lunches with the students each day, encouraging conversation even during leisure time.

Thanks to California's liberal school accreditation laws, Brightworks will be able to hand out diplomas, despite the fact that it will do little to prepare its students for the SATs. The school states up front to parents that if they expect their children to take the SATs, they can enroll in an afterschool SAT prep program. Tulley says the school administration has approached a few colleges and explained their curriculum, asking whether a student's completed portfolio would be considered for admission, and received uniformly affirmative responses.

Tulley admits he has a lot riding on Brightworks. "I just refinanced my house," he tells me. "Everything is on the line."

Indeed, everything is on the line, and not just at Tulley's new school. As this book was being completed, the national unemployment rate in the United States was hovering at around 8.2 percent, well above the level considered as "full employment" of 5 percent. But at the same time, small business owners and recruiters reported having difficulty filling existing jobs. Most claimed they couldn't find enough workers with the technical skills the jobs required. Among the toughest positions to fill were those for software and information technology personnel, mechanics, and machine operators. Number one on the list, according to one survey, was engineers.

Only time will tell if America is on the road to recovering its tinkering spirit, but there is no doubt that more people than ever are devising new ways to rekindle the spark. The issue has even trickled down into popular culture via some unlikely sources.

I recently happened upon a graphic novel titled *Tinkerers: An Original Tale of the New Future*, written by the respected science fiction author David Brin, best known for his Uplift Universe series, and published by the Metals Service Center Institute, a trade association "that represents about 275 companies that make and distribute industrial metals," in 2010.

Despite the obvious bias of the publisher, the book has its virtues. In a brightly rendered comic format, Brin cleverly envisions a not so distant future in which jetpacks and hover cars are commonplace but America's infrastructure is crumbling. The evocative artwork, by Jan Feindt, expertly contrasts the glamorous technology that surrounds us with a declining society rotting at its core.

The protagonist, Danny Nakamura, and his high school class are nearly killed on graduation day, when an old bridge collapses. Nakamura subsequently goes on a journey to find out what went wrong in America. Some of the sage characters he meets along the way offer some possible explanations. "Americans are the world's teenagers," says one. "Her virtues and her faults were always those of adolescence . . . like our quick enthusiasm and easy boredom." Another suggests that the United States triumphed in World War II not only due to its courage and productivity, but also due to its army of tinkerers, resourceful young men who grew up playing with cars and radios, who expertly maintained the machines needed to fight the noble war. In contrast, without a unified cause, today's young men spend their days mastering video games instead.

Some other reasons floated for the envisioned American decline include the cult of individualism, an educational system that moved away from discipline and memorization just to make learning more interesting, banks run by MBAs who weren't really bankers, and excessive military spending.

By the end, the bridge that collapsed at the beginning is rebuilt by small teams of separately trained experts who practiced their roles through simulations and communicated with each other throughout the process. "It took new methods of design, distributed-fabrication, webbed-integration, agile finance," says Nakamura. "And yet . . . my quest had taught me the crucial ingredients were human: Pride in nation and community; love of progress, civilization; curiosity and craft; and a world whose damage can be healed by caring skill."

Charmingly cheesy and a bit heavy-handed at points, Brin's *Tinkerers* somehow captures a bit of the tone that the United States likely needs to adopt to get its tinkering groove back. After all, a strange combination of seriousness and frivolity has always served us well. We are the world's teenagers, in more ways than we care to admit. But even the most brilliant teenager sometimes needs to take a break from the grind, and take a breath to contemplate his or her promising future.

CHAPTER **11**

CONCLUDING THOUGHTS
ON TINKERING

THIS BOOK HAS IDENTIFIED THE PROCESSES and thought patterns intrinsic to a uniquely American style of tinkering. It was
born out of the idea that Americans were losing sight of a key trait
that helped make us a great country in the first place. My hope is
that the stories contained in these pages will inspire readers to think
about how they might incorporate a tinkering mindset into their
own lives, as well as their children's.

I deliberately have avoided filling this book with prescriptions for
educational reform or remedies for reordering America's priorities
and values, because I think such assessments tend to be wildly sub-
jective and scolding. My hope is that this fairly orderly, if occasionally

random, survey of America's tinkering history, proudly dilettantish but equally passionate in its pursuit of something useful, will provide examples of how people might discover their inner tinkerer or support the tinkering spirit in someone they know.

I want to emphasize that reviving the American tinkering spirit does not simply mean graduating more science majors and engineers, as Dean Kamen suggests in Chapter 3. While there's no doubt that much of the innovation occurring today is in the realm of technology, I believe that a society overly focused on acquiring specialized technical skills, or indeed specialized skills of any kind, loses the ability to think big. I worry that we have become a nation of specialists, the enemies of true tinkering.

This may be a function of the rising cost of higher education. As college tuitions have skyrocketed—at an average annual rate of around 8 percent, about twice the rate of inflation, meaning the cost of college doubles every nine years—educational experts have focused more and more on the absolute return on investment, rather than on the value of a well-rounded thinker who can figure out how to tinker based on a broad understanding of the world as a whole. If parents are going to be forking over a couple of hundred grand to educate their offspring, the logic goes, there should be a clear payoff, usually defined as a high-paying position in a rapidly growing industry.

Objecting to this perspective might strike some as cultural elitism. After all, who but the most privileged members of our society can afford to shell out hundreds of thousands of dollars with only the vaguest hope that their children will be able to fend for themselves in the increasingly rocky global economy? But the starkly practical approach to education may not be the guarantee for long-term success that it appears on the surface. While a 2007 study by NFI Research reported that more than half of the senior executives and managers surveyed said their organizations favored specialists over generalists, it also reported that generalists were favored in around a third of organizations, and 20 percent said their departments would be more effective with more generalists than specialists.

In other words, managers preferred their subordinates to be specialists but their colleagues to be generalists. "The irony of corporate America is that while generalists drive innovation and long-term results, specialists are most often rewarded at the vice president level and below," explained one survey participant.

Meanwhile, scholarships and other forms of financial aid have put a classic liberal arts education within the reach of more Americans. But if the virtues of being a generalist are obscured by society's so called pragmatists, such opportunities are likely to be squandered at the expense of future tinkering. If your role in society is largely predetermined by the kind of education you have received and the career you've cleverly staked out, the likelihood of happening upon some new, unlikely combination of existing elements and thereby transforming an aspect of the way the world works is greatly reduced.

This focus on specialization also lacks that distinctly American belief that anything is possible if you put your mind and best effort behind it. A big part of the American tinkering spirit is about finding inspiration in the creative pocket that exists between the metronomic beats of business as usual. That American style of seeing possibility where others see nothing is why people like Steve Jobs and Warren Buffett have become contemporary folk heroes. In recent years, in apparent acknowledgment of this tendency, job recruiters have begun seeking what are termed "T-shaped individuals"—with the vertical bar referring to deep expertise in a specific discipline and the horizontal bar representing an ability to work with experts from other disciplines—who exhibit understanding in areas beyond their specialty.

The jazz critic Gary Giddins once wrote of Louis Armstrong that few fans of the legendary jazz trumpeter and entertainer, known for his virtuosic improvisational abilities and genre-busting experimentation (musical tinkering?) as much as for his comedic stage persona, understood how truly influential he was: "How many of those who enjoyed Benny Goodman's swing caravan in the thirties or rocked to Chuck Berry in the fifties or savored the increased vibrato that became fashionable in the brass sections of symphony orchestras knew the extent to which they were living in a world created by the famous gravel-mouthed clown?"

True American tinkerers are like Louis Armstrong. They operate within an existing vernacular and yet break new ground by the sheer force of their creativity and exuberance. Certainly there are those operating in the scientific and engineering fields who meet these criteria, but to limit the role of tinkering to their efforts alone would be nothing short of foolish. While not everyone is a brilliant tinkerer, everyone has the ability to be creative.

In their misguided attempt to become more results oriented, many American school systems have become more focused on raising test scores than on immersive, process-based learning that incorporates some of the ideas put forth by Gever Tulley's Tinkering School.

Some are trying to correct the effects of this trend at the graduate level, In 2005, David Kelley, founder of IDEO, a renowned innovation and design firm based in Palo Alto, California, established the Hasso Plattner Institute of Design at Stanford University, a nondegree program known as the d.school, to teach students in the university's seven graduate schools how to embrace ambiguity, experimentation, and the possibility of failure. George Kemble, the d.school's executive director, told the *Wall Street Journal* in 2011 that much of what the d.school did was help students unlearn what they learned in elementary school. Kemble explained that a fear of failure was very common among students accustomed to taking standardized tests. "What we want the graduate students to do is work with others and go out and take risks," he said.

My own daughter's experience one year in elementary school was, I worry, far from atypical. We live in a Connecticut suburb not far from New York City known for its exceptional public schools. It's the kind of town families move to primarily because the schools are good and well run. Generally, the caliber of teachers and students in these kinds of school systems matches those of the nation's best private schools.

But one year, in particular, my daughter had the misfortune of being placed in the class of one of her school's less-inspiring educators. Whereas most talented teachers know instinctively that "teaching to the test" is a bare-minimum requirement that must be augmented with other materials and creativity, this teacher was adamant about sticking firmly to the curriculum as stipulated by the school board and by state requirements.

Even worse, the teacher was maddeningly literal in her interpretation of her students' performance—she once excoriated our daughter for decorating a penmanship exercise with color markers after she had completed the assignment, scrawling "Be Neater!" across the top of the page—and overzealous in her use of discipline and scolding as a motivator. Beyond those qualities, she expressed little interest in topics that didn't pertain directly to the day's lesson, and made no effort to incorporate current events into her class's discursive flow.

For that whole school year, my wife and I gritted our teeth and tried our best to shield our daughter from what we couldn't help but judge to be a harmful influence on our daughter's educational development. During our family time, we sought to augment our daughter's classroom experiences with other, more stimulating experiences: favorite books, museums, theater, word puzzles, and math workbooks. She has always been artistically inclined, so we enrolled her in after-school art clubs and weekly piano lessons.

After some prompting from us during a parent-teacher conference, the teacher offered our daughter the opportunity she had been waiting for: the chance to develop an optional project she would later present to her classmates. My daughter was visibly nervous, but also quite excited. Finally, she would have an opportunity to highlight her individual interests and creativity.

A budding art fan, she chose Pablo Picasso as the subject of her presentation. My wife and I couldn't have been more pleased. We enthusiastically took her on a trip to the Museum of Modern Art in New York, where she studied original Picassos close up, and chose one to focus on in her presentation. Her ambitious plan was to explain cubism to her classmates, what they were looking at, and why Picasso painted the way he did. She began her research online and dutifully wrote out key points on index cards to coalesce and organize her thoughts.

Unfortunately, her teacher seemed to have forgotten all about the assignment by the next week and made no effort to follow up. My daughter was unbowed, however, and continued developing her ideas, even without her teacher's input. She made her own rendition of a Picasso-style painting and combed through a few art books she checked out from the local library.

After a round of parental intervention, the project gained steam again and finally was scheduled. By all accounts, it went off very well. However, at the end of the term, there was no mention of the project on her standardized report card (alas, these days, even the comments on report cards are often standardized, chosen from a prewritten list).

My wife and I gave a collective sigh of relief once the year was over. One moment during the first few weeks of summer, after enough time had passed since the end of the school year, my daughter off-handedly confessed that one of her favorite experiences during the school year had been packing up the classroom at the end of the spring term. She explained how she enjoyed figuring out where things could fit and how she gained a sense of accomplishment once everything was finally put away. "It's more fun than just sitting at a desk all the time," she told me.

Over the past few years, a new kind of company has emerged in the United States that seems to acknowledge that preaching the importance of learning more practical skills is not a very good way of encouraging tinkering or the innovation that it produces. And though each company does it somewhat differently, the basic idea is the same. Rather than attacking the problem from the front end and putting pressure on the tinkerers, these companies address the back end by, in effect, financing tinkering that seems to exhibit some degree of promise but that for whatever reason has not found backers or a traditional support system to help it flourish.

Many of these companies conduct business through a model known as "crowdfunding." Crowdfunding harnesses the interactive power of the Internet to convince strangers to donate money to sponsor a particular invention or creative project and then provides them with a reward once the project is completed. The best-known crowdfunding company at the time of this writing is Kickstarter in New York.

Another related business model is known as a "seed accelerator," a reference to its rapid approach to seeding, or funding, start-up technology companies. Unlike the traditional world of venture capital firms, in which a large number of entrepreneurs compete for a relatively small number of lucrative funding opportunities, seed accelerators create a framework that provides smaller funding opportunities to a much larger group of start-ups, as well as a support organization that provides advice and contacts to fledgling companies. Y Combinator, based in Mountain View, California, is probably the most established. Another is TechStars, out of Boulder, Colorado.

Y Combinator, cofounded in 2005 by Paul Graham, a former computer programmer with a PhD from Harvard, organizes itself as a boot camp for start-ups, holding two three-month-long sessions a year, during which fifteen to twenty entrepreneurs learn the ropes from Y Combinator's experienced pros in exchange for 6 percent of the resulting company's equity. Each start-up is given a relatively small amount of capital, usually between $14,000 and $20,000, depending on the number of founders, just enough to allow participants not to work additional jobs while participating in the programs. Some of the more than three hundred companies Y Combinator has helped launch have gone on to great success, including Airbnb, Dropbox, and Scribd.

TechStars, started by serial entrepreneur David Cohen, takes more of an *American Idol* approach, auditioning more than one thousand applicants for each of its start-up programs in Boulder, New York City, Seattle, Boston, and San Antonio, and ultimately accepting less than one percent. Its spring 2012 event in New York had fifteen hundred applicants, of which fourteen were chosen. TechStars provides winners with $14,000, expert guidance, unlimited access to a network of technology mentors, as well as free office space, also in exchange for a 6 percent equity stake in each company. SendGrid, Sensobi, and Filtrbox are a few of the resultant concerns.

While Kickstarter is adamant about its role as a booster of artistic projects rather than businesses, it comes closest in my mind to playing the role of a tinkering incubator. Maybe that's because its founders have the kinds of generalist backgrounds that most resemble those of the tinkerers being undersupported by the contemporary American business community.

Kickstarter was the brainchild of cofounder Perry Chen, also the company's chief executive. While living in New Orleans in 2002, he decided he wanted to organize an electronic-music concert featuring Austrian DJs Kruder and Dorfmeister. The show never came together, as Chen quickly realized it required too much personal financial risk. He wondered whether he could create a web application that would help raise money for such an event from others, thus lessening the burden on the individual innovator.

Chen didn't make much progress on his idea for the next three years, but he didn't stop thinking about it. He returned home to New York in 2005, and found work as a waiter. One day he found himself describing his concept to a customer, Yancey Strickler, a young rock music journalist. Strickler was intrigued, and the two began devising a plan of action. Together with a third partner, Charles Adler, who brought technological knowhow to the table, Chen and Strickler struggled to make Kickstarter a reality. They nearly gave up more than a few times over the next few years, but Chen persisted. Kickstarter officially launched in April 2009.

Kickstarter's selection process for each project it sponsors is something apart from the traditional route followed by tinkerers. Applicants pitch their idea on Kickstarter's website and must choose a specific amount of funding they need to complete it and a deadline for raising the money. They also must offer a reward to participants, anything from a personalized thank-you note, a credit on an album cover, or a customized T-shirt. The typical donor gives a small amount, commonly $25.

If the project fails to generate the dollar amount requested by the deadline, it doesn't move forward. If it raises enough money in time, the project goes ahead, financed by the group of small donors.

Not any project can get funding via Kickstarter. The company employs a staff of screeners to filter out proposals that don't fit their concept of supportable endeavors. In general, Kickstarter won't sanction charities, political causes, careers, or start-ups—these goals are apart from pure creativity and receive support elsewhere—nor will it approve projects that consist of nothing more than a request for handouts to buy something. The company encourages participants to craft clever pitches for their idea, which frequently include homemade videos.

One of the earliest successful Kickstarter campaigns was a project pitched by Allison Weiss, an Atlanta-based musician, who wanted to raise $2,000 in sixty days to fund the recording of her latest CD. Weiss raised that amount in only ten hours and eventually received $7,711 from more than two hundred donors.

Kickstarter is eager to keep its mission focused on creative works, as opposed to gadgets and businesses, but it's been something of a struggle. One of the reasons is that the concept has caught on so rapidly—Kickstarter was on track to raise around $300 million in funding in 2012, three times the amount it raised in 2011—that the amounts being raised have escalated beyond the modest art project proportions the cofounders first envisioned. As example, Amanda Palmer, a musician once signed to an independent record label, launched a Kickstarter campaign in 2012 to raise $100,000 in thirty days to complete and promote her new album. Instead, in a month's time, she collected $1,192,793 from 24,883 people.

The other reason is that, while Kickstarter still funds plenty of music, film, and art projects, it also has succeeded in funding some high-profile products resulting from tinkering. The best-known example is the Pebble, a customizable electronic paper watch with a display similar to that of an Amazon Kindle e-book. In classic tinkering style, the prototype of the Pebble was built from spare cell-phone parts. Pebble's creator, Eric Migicovsky, launched a campaign for $100,000 in May 2012; he ultimately raised more than $10 million from nearly 70,000 backers.

But the essence of the Kickstarter process has remained largely intact. According to a source I spoke with close to the company, the cofounders view themselves as modern-day Medicis, reinventing the patronage model on democratic terms. That means that Kickstarter backs projects big and small, both mainstream and quirky, with equal enthusiasm. Because the company extracts a 5 percent fee from the total amount collected for each project, it certainly benefits from projects that achieve seven- or eight-figure donations. But since Kickstarter doesn't take any equity in those projects, it quickly and happily moves on to the latest and greatest proposals.

Among the more serendipitous tinkering projects that have been funded through Kickstarter are a line of dress shirts that use technology developed at NASA to control perspiration, reduce odor, and eliminate wrinkles; a stainless steel coffee bean that prevents a cup of coffee from getting too hot or too cold; the C-Loop, a camera strap that attaches to the tripod mount on the bottom of the camera rather than the hooks on the top to prevent the strap from getting in the way of the lens; and an innovative textile printing process that uses sunlight to develop images.

The beauty of the Kickstarter approach is that the would-be patrons that sign on to fund a project become part of the story of the project's evolution. Each concept lives or dies based on the direct interest of a relatively random group of observers. By tapping into this untethered enthusiasm, tinkerers become immersed in a very free-form process of discovery that almost miraculously removes them from the corporate ecosystem that can be so stifling.

There's some irony in the fact that technology has helped return tinkering to a state not that far off from the way Benjamin Franklin must have experienced it: unpredictable, unencumbered, and swirling with possibility. I suspect that America's tinkering spirit is a cyclical resource, prone to fallow and fertile periods that are impacted at times by major world events, such as World War II. And for all the disruptive power that tinkerers hold, theirs is ultimately a noble cause. After all, tinkering rescues us from a far riskier fate, that of stagnation.

As long as the United States continues to make room for and accommodate those who see things differently and remain determined to make their visions prevail against all odds, tinkering will remain our most precious, but renewable, natural resource. Whether that tinkering is physical or virtual matters less than the level of freedom and space given to those who practice it. Knowing that it is okay— indeed, necessary—to go beyond what is recommended or permitted by an authority, whether pedagogical or commercial, is intrinsic to the American tinkering spirit. Being difficult yet diligent, determined yet daydreamy—somewhere within these kinds of contradictions lies the future of our national connectedness. All that stands between here and there is cockeyed bravery.

ACKNOWLEDGMENTS

Simply put, this book itself required a lot of tinkering.

The story of its creation is, in one regard, the tale of two Tims. It began with a conversation that my agent, Paul Bresnick, had with the book's original editor, Tim Sullivan. Tim had an idea for a book called *The Tinkerers,* inspired by Daniel J. Boorstin's Knowledge Trilogy (*The Discoverers, The Creators, The Seekers*). From that point, I ran with the idea and fleshed out what that book might be. I give Tim Sullivan a heap of credit for saying yes to nearly everything I suggested. And thanks to Paul Bresnick for making it all happen.

The first Tim eventually moved on from Basic Books. My new editor, Tim Bartlett, thankfully had passion for the project and added his own enthusiasm into the mix. Tim Bartlett proved to be an expert collaborator and a pitch-perfect sounding board. His patient and careful editing of the text made nearly every sentence stronger. I am grateful for his ongoing encouragement and engaged participation. Additional thanks to his assistant, Sarah Rosenthal.

I also owe thanks to all of the contemporary tinkerers who agreed to be interviewed for this book. I had a pretty specific notion of what I hoped to get out of them and none complained, not even once, when I persisted in trying to get it.

My wife, Erica, was a true collaborator on this project, as well. She helped me carve out the time from a hectic family schedule to focus on the work and assured me at every step along the way that it was worth it.

Last, I'd like to thank my children, Charlotte and Henry, whose intelligence and curiosity partly inspired this book. Their natural inquisitiveness and interest in figuring out how stuff works helped me appreciate that the American tinkering spirit lives within every new generation, just waiting to be awakened.

NOTES

CHAPTER 1: WISING UP ABOUT A SMARTPHONE

3   **the remarkable case of George Hotz:** "Machine Politics," by David Kushner, *New Yorker*, May 7, 2012, pp. 24–30.

9   **earn degrees in science or engineering:** "The Electrifying Edison," by Bryan Walsh, *Time*, July 5, 2010.

9   **50.7 percent of new patent grants:** "Ben Franklin, Where Are You?" by Michael Arndt, *Bloomberg BusinessWeek*, December 28, 2009 and January 4, 2010, p. 29.

10   **as far into the future as you can imagine:** *Conversations with Leading Economists: Interpreting Modern Macroeconomics* by Brian Snowden and Howard R. Vane (Edward Elgar Publishing, 1999), p. 310.

11   **dumped into the Gulf every four days:** "The Poisoning," by Jeff Goodell, *Rolling Stone*, August 5, 2010.

12   **sensed a sudden change in pressure:** "Robots Working 5,000 Feet Underwater to Stop Flow of Oil in Gulf of Mexico," by Campbell Robertson and Clifford Krauss, *New York Times*, April 26, 2010.

13   **titled "America Goes Dark":** "America Goes Dark," by Paul Krugman, *New York Times*, August 8, 2010.

17   **according to United Nations statistics:** "Despite China's Might, U.S. Factories Maintain Edge," by Paul Wiseman, Associated Press, February 1, 2011.

CHAPTER 2: TINKERING AT THE BIRTH OF A NATION AND BEYOND

20   **dismiss him as mere tinkerer:** *Benjamin Franklin: An American Life* by Walter Isaacson (Simon & Schuster, 2003), p. 129.

21   **" . . . the ruts their fathers trod":** *George Washington: Farmer: Being an Account of His Home Life and Agricultural Activities* by Paul Leland Haworth (Bobbs Merrill Company, 1915), p. 6.

22   **"and found She answerd very well":** *George Washington: Farmer* by Paul Leland Haworth (2004–03–01). (Kindle location 739), public domain books, Kindle edition.

22   **Potomac River as a route for commerce:** *The Grand Idea: George Washington's Potomac and the Race to the West* by Joel Achenbach (Simon & Schuster, 2004), p. 129.

23   **limiting the value of their knowledge:** *Washington: The Indispensable Man* by James Thomas Flexner (Little Brown, 1974), p. 197.

24   **had ever seen a canal lock before:** *Patowmack Company Canal and Locks* by Ricardo Torres-Reyes (Division of History, Office of Archaeology and Historic Preservation, U.S. Department of the Interior: National Park Service, May 1, 1970).

26   **printed the** *American Weekly Mercury***:** *Benjamin Franklin: An American Life*, p. 113.

27   **job with William Hunter of Virginia:** Ibid., p. 157.

30   **" . . . two, three weeks, a month":** *Divided Highways* by Tom Lewis (Viking Penguin, 1997), p. 6.

CHAPTER 3: CONTEMPORARY TINKERER FINDS HIS WAY

39   **tinkerer traits at an early age:** *Reinventing the Wheel: A Story of Genius, Innovation, and Grand Ambition* by Steve Kemper (HarperBusiness, 2003), p. 9.

40   **particularly ones called thyristors:** "The Big Deal: Inventor Dean Kamen," by Victoria Barret, Forbes.com, March 31, 2010.

57   **being made in American innovation:** *Race Against the Machine: How the Digital Revolution Is Accelerating Innovation, Driving Productivity,*

and Irreversibly Transforming Employment and the Economy by Erik Brynjolfsson and Andrew McAfee (Digital Frontier Press, 2011).

CHAPTER 4: EDISON'S FOLLY REINVENTS TINKERING FOR THE MODERN AGE

62 **telegraph communities in the nation:** *Edison: A Life of Invention* by Paul Israel (John Wiley & Sons, 1998), p. 40.

64 **according to biographer Randall Stross:** *The Wizard of Menlo Park: How Thomas Alva Edison Invented the Modern World* by Randall E. Stross (Crown, 2008), p. 13.

70 **" . . . would start sawing," he explained:** *Edison, His Life and Inventions* by Frank Lewis Dyer (Harper & Bros., 1910).

72 **superior manual coordination to operate:** *Edison: Inventing the Century* by Neil Baldwin (Hyperion, 1995), p. 82.

77 **tinkering for the contemporary era:** *Edison: A Life of Invention*, p. 167.

81 **protect the country's interests internationally:** *Soldiers of Reason: The Rand Corporation and the Rise of the American Empire* by Alex Abella, (Houghton Mifflin Harcourt, 2008) p. 13.

81 **committed itself to solve:** Ibid., p. 54.

83 **" . . . What kind of payload?":** Ibid., p. 58.

86 **" . . . face the consequences of failure":** "Robert S. McNamara, Architect of a Futile War, Dies at 93," by Tim Weiner, *New York Times*, July 7, 2009.

CHAPTER 5: MYHRVOLD'S MAGIC TINKERING FACTORY

91 **geophysics and space physics:** *The Microsoft Way: The Real Story of How the Company Outsmarts Its Competition* by Randall E. Stross (Basic Books, 1997), p. 54.

93 **an offshoot of Intellectual Ventures:** "Billionaire Nathan Myhrvold's $625 Cookbook," *Bloomberg Businessweek*, November 11, 2010.

100 **"being used to sue companies that do":** "When Patents Attack," National Public Radio, *All Things Considered*, July 26, 2011 broadcast.

102 **returns it had registered so far:** "Trolling for Suckers," by Nathan Vardi, *Forbes*, August 8, 2011.

102 "... the *Harvard Business Review*": Funding Eureka!" by Nathan Myhrvold, *Harvard Business Review*, March 2010.

CHAPTER 6: WHEN TINKERING VEERS OFF COURSE

107 **more promiscuously than ever:** *The Rational Optimist: How Prosperity Evolves,* by Matt Ridley (Harper, 2010) pp. 6, 352.

108 **turbine plant in Schenectady, New York:** "Remarks by the President on the Economy in Schenectady, New York," whitehouse.gov, January 21, 2011.

110 **the "Morgan mafia":** "The Dream Machine," by Gillian Tett, *Financial Times*, March 25, 2006.

113 **making commercial loans:** "The $58 Trillion Elephant in the Room," by Jesse Eisinger, Portfolio.com, October 15, 2008.

CHAPTER 7: THE TINKERER ARCHETYPE IS REBORN

123 **the country's main export:** "Climate Change and the End of Australia," by Jeff Goodell, *Rolling Stone*, October 13, 2011, p. 57.

129 **many of them children:** "A Life of Its Own: Where Will Synthetic Biology Lead Us?" by Michael Specter, *New Yorker*, September 28, 2009, p. 56.

134 **the Boston Consulting Group:** "Google Tries Something Retro: Made in the U.S.A.," by John Markoff, *New York Times*, June 27, 2012.

CHAPTER 8: PARC AND THE POWER OF THE GROUP

138 **most successful industrial product in history:** *Dealers of Lightning: Xerox PARC and the Dawn of the Computer Age* by Michael A. Hiltzik (HarperCollins, 2004), p. 22.

140 **computer science research program:** "Space War," by Stewart Brand, *Rolling Stone*, December 7, 1972, p. 52.

147 **rather than an open one:** *Open Innovation: The New Imperative for Creating and Profiting from Technology* by Henry Chesbrough (University of Oxford Press, 2008) p. 5.

CHAPTER 9: A TRIO OF ALTERNATIVE TINKERING APPROACHES

160 **now known as Rovio, on his own:** "How Rovio Made Angry Birds a Winner (and What's Next)," by Tom Cheshire, *Wired*, April 2011.

165 **in less than an hour:** "Jeanne Gang: The Art of Nesting," by Stephen Zacks, *Metropolis*, June 2008.

CHAPTER 10: A DIFFERENT KIND OF SCHOOL

184 **difficulty filling existing jobs:** "A Sea of Job-Seekers, but Some Companies Aren't Getting Any Bites," by Darren Dahl, *New York Times*, June 27, 2012.

CHAPTER 11: CONCLUDING THOUGHTS ON TINKERING

188 **doubles every nine years:** http://www.finaid.org/savings/tuition -inflation.phtml.

189 **more generalists than specialists:** "Specialists vs. Generalists," by Chuck Martin, *CIO Magazine*, April 5, 2007.

190 **" . . . famous gravel-mouthed clown?":** *Satchmo: The Genius of Louis Armstrong* by Gary Giddins (Da Capo, 2001), p. 6.

190 **" . . . take risks," he said:** "Innovation 101," by Carolyn T. Geer, *Wall Street Journal*, October 16, 2011.

196 **more than two hundred donors:** "The Trivialities and Transcendence of Kickstarter," by Rob Walker, *New York Times*, August 7, 2011.

INDEX